W.W Fisher

A Class Book of Elementary Chemistry

W.W Fisher

A Class Book of Elementary Chemistry

ISBN/EAN: 9783337062408

Printed in Europe, USA, Canada, Australia, Japan

Cover: Foto ©berggeist007 / pixelio.de

More available books at **www.hansebooks.com**

Clarendon Press Series

A CLASS BOOK

OF

ELEMENTARY CHEMISTRY

BY

W. W. FISHER, M.A., F.C.S.

ALDRICHIAN DEMONSTRATOR OF CHEMISTRY
FORMERLY FELLOW OF CORPUS CHRISTI COLLEGE

With Sixty Engravings on Wood

Oxford

AT THE CLARENDON PRESS

M DCCC LXXXVIII

PREFACE.

AS the teaching of the Elements of Chemistry is becoming more commonly a part of regular school work, and this subject is included in nearly every course of liberal education, it is hoped that this Class Book will be found useful as a compendium of the fundamental laws, principles and facts of Chemical Science.

My purpose throughout has been to give as briefly as possible some account of the most important chemical phænomena, actions and changes, with the laws of chemical combination and the theoretical explanations of those laws commonly accepted.

After a necessary introduction, in which a general outline of chemical phenomena and the laws of chemical action is given, in the earlier chapters Water and Air are treated with some detail. Since they are themselves bodies of the greatest importance, a more complete treatment of their chemical and physical properties and characters is adopted than would be possible with the rest of the subject-matter. The experimental demonstrations introduced serve as a general introduction to chemical method and operations.

It is presumed that the Class Book will be used only as an adjunct to experimental lectures and other practical work; but illustrations with explanations of important experiments are furnished to make the work more useful and complete. In many cases however experimental verifications of facts given in the text are merely suggested, without detailed descriptions being given.

In the succeeding chapters the Elements, Carbon, Sulphur, Chlorine, Nitrogen and Phosphorus, and also Fluorine, Boron and Silicon are taken in succession, with their chief compounds and the important acids derived from them. A chapter on the Periodic Law, with a sketch of the theories relating to Acids, Bases and Salts is next introduced, as the student at this stage will be able to compare and classify the characters of the non-metallic elements and in consequence will be in a better position to undertake the study of the metallic elements, of which the distinctive characters are much less easily grasped.

The metals are treated simply in outline and only their important and characteristic compounds touched upon. A classification based upon the Periodic Law has been adopted. Although students can only attain a satisfactory knowledge of this branch of chemistry by actual work at the Laboratory bench it is hoped that the outlines of the metals here given will be found useful as an introduction to their study. The recent researches by V. Meyer, Scott, and others, upon the vapour densities of the chlorides of Aluminium, Iron, etc., having fully established the simplest molecular constitution of these bodies, their formulae have been modified, and in a few instances therefore are not in accordance with those in recent use.

In the selection of subjects I may state that I have followed in the main the syllabus of the Oxford Local Examinations for Senior Candidates and the Examination of Women, which is similar in extent to the syllabus of the Preliminary Examination in the School of Natural Science and the Preliminary Examination for Medicine at Oxford. The illustrations and diagrams are for the most part drawn by myself from the apparatus I have been in the habit of using in lectures or from the original diagrams in the memoirs of Dumas and Boussingault on the composition of Air, Water, and Carbon Dioxide; but my thanks are due to the Delegates of the Press, and to Professor Williamson and Messrs. Harcourt and Madan, for permission to introduce some other blocks which are also employed.

I have likewise to record my thanks to my colleague, Mr. J. E. Marsh of Balliol College, for kindly assisting in the revision of proof sheets.

W. W. FISHER.

OXFORD: August 13, 1888.

TABLE OF CONTENTS.

Chapter I.

Elements and Compounds. Laws of Chemical Combination. Atoms and Molecules. Atomic Weights. Formulae . 1

Chapter II.

Hydrogen. Decomposition of Water 14

Chapter III.

Oxygen. Ozone 21

Chapter IV.

Water. Synthesis of Water. Composition by Weight and Volume. Latent Heat of Water and Steam. Solutions. Natural Waters. Distilled Water 28

Chapter V.

Nitrogen and Air. Composition of Air by Weight and by Volume 42

Chapter VI.

Carbon. Forms of Carbon; Carbon Oxides, Combustion, and Flame 50

Chapter VII.

Sulphur; Sulphur Oxides, Sulphurous Acid, Sulphuric Acid; Hydrogen and Sulphur . . . .

Chapter VIII.

The Halogen Elements. Hydrochloric Acid; Chlorine, Bromine, and Iodine . . . .

Chapter IX.

Nitrogen; Ammonia and Nitric Acid . . .

Chapter X.

Oxides of Nitrogen. Hydroxylamine. Diamine .

Chapter XI.

Phosphorus. Phosphorus Chlorides. Phosphorous Acid. Phosphoric Acids. Phosphine . . . .

Chapter XII.

Oxidized compounds of Chlorine, Bromine and Iodine. Fluorine . . . .

Chapter XIII.

Boron and Silicon. Silica, Silicates, and Glass . . .

Chapter XIV.

Arsenic . . . .

Chapter XV.

Quantivalence and the Periodic Law. Acids, Bases, and Salts. Calculation of Formulae . . . .

Table of Contents.

CHAPTER XVI.
Metals. The Alkali Metals Potassium and Sodium. The Alkali manufacture 156

CHAPTER XVII.
Ammonium Salts 168

CHAPTER XVIII.
Dyad Metals; the Alkaline earth metals Calcium, Strontium, and Barium 172

CHAPTER XIX.
Dyad Metals; Beryllium, Magnesium, Zinc, and Cadmium . 181

CHAPTER XX.
Triad Metals; Aluminium. Clay, Pottery, and Porcelain . 188

CHAPTER XXI.
Iron, Nickel and Cobalt. The Blast Furnace. Production of Cast Iron, Wrought Iron, and Steel 193

CHAPTER XXII.
Hexad Metals; Chromium and Chromates 206

CHAPTER XXIII.
Heptad Metals; Manganese, Manganates, and Permanganates 210

CHAPTER XXIV.
Metals of the Copper Group; Copper, Silver, and Mercury . 214

CHAPTER XXV.
Tetrad Metals; Tin and Lead 231

Table of Contents.

Chapter XXVI.

Pentad Metals; Antimony and Bismuth . . . 241

Chapter XXVII.

Atomic Weights and Specific Heats; Atomic Heat. Molecular Heat 248

Chapter XXVIII.

The Physical properties of Gases. Liquefaction. Diffusion. Expansion by Heat. Boyle's law 253

Index 267

THE METRIC SYSTEM.

The Decimal system of measures of length, area, volume and weight, used in France and Germany, and employed for reasons of convenience by scientific men in this country, is based upon the standard METRE.

All measures of length are referred to this standard, which is equivalent in length to the ten-millionth of the earth's quadrant. The standard metre is made of platinum, and is deposited in the archives at Paris; copies of it have been constructed and carefully adjusted so as to be equal in length. In comparison with English measures the Metre equals

1·0936 yards,
or   3·2809 feet,
or   39·37079 inches.

Measures of Length.

The decimal multiples and decimal fractions of the Metre are indicated by Greek or Latin prefixes. Thus δέκα 10, ἑκατόν 100 and χίλια 1000 give the multiples decametre (10 m.), hectometre (100 m.), and kilometre (1000 m.); while from the Latin *decem* 10, *centum* 100, *mille* 1000, are derived decimetre (·1 m.), centimetre

(.01 m.), and millimetre (.001 m.). The measures most commonly used in this country are given in the following tables.

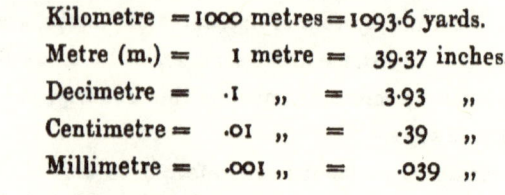

Kilometre = 1000 metres = 1093.6 yards.
Metre (m.) = 1 metre = 39.37 inches.
Decimetre = .1 ,, = 3.93 ,,
Centimetre = .01 ,, = .39 ,,
Millimetre = .001 ,, = .039 ,,

*** A kilometre is nearly 1100 yards, and 8 kilometres are roughly a little less than 5 miles.

The metre is nearly 40 inches, and the decimetre, as will be seen from the figure, is nearly 4 inches. Also a millimetre is approximately $.04 = \frac{1}{25}$ inch.

$$1 \text{ inch} = 2.54 \text{ c.m.}$$
$$12 \text{ inches} = 1 \text{ foot} = 30.48 \text{ c.m.}$$

Measures of Volume.

If we take a cube, of which the side is one decimetre or 10 centimetres, we obtain a unit volume of one cubic decimetre known as one litre; this is 1000 cubic centimetres.

1 litre or cubic decimetre = 1000 c.c. (cubic centimetres) and is nearly 1.76 pints.

Inches and Centimetres.

*** The litre and the cubic centimetre are the units commonly employed in the laboratory for measuring volumes of liquids or gases. The Hectolitre (100 litres) is used as a unit of measure for wine, oil, etc., for trade purposes.

$$1 \text{ gallon} = 4 \text{ quarts} = 4.544 \text{ litres.}$$
$$1 \text{ pint} = 20 \text{ fluid ozs.} = 567.9 \text{ c.c.}$$

Measures of Weight.

The unit mass employed for weighing is the **gramme**. A cubic centimetre of water at the temperature of 4°C., at which the water is at its maximum density, weighs one gramme. The kilogramme (1000 g.) is the only multiple in common use, the decimal fractions used being the decigramme, the centigramme, and the milligramme.

1 kilogramme	= 1000	grammes	=	2.205 lbs. (avoird.)	
1 gramme	=	1 gramme	=	15.43 grains.	
1 decigramme	=	.1 ,,	=	1.54	,,
1 centigramme	=	.01 ,,	=	.154	,,
1 milligramme	=	.001 ,,	=	.015	,,

∗∗∗ The volume of one gramme of water is one cubic centimetre; and the volume of a kilogramme of water is one litre (1000 c.c.).

1 grain	=	.0648	gramme
7000 grains = 1 lb. avoirdupois	= 453.6	,,
1 oz. ,,	= 28.35	,,

ELEMENTARY CHEMISTRY.

CHAPTER I.

It is a familiar fact that the world in which we live and the objects around us are composed of an infinite variety of material substances of the greatest diversity of characters. The rocks composing the Earth, such as granite, sandstone, limestone, chalk, slate and clay; the minerals and ores from which metals are produced, are material substances obviously different from each other in many respects; while in the world of living beings the varieties of animals and plants and the substances formed in the animal and vegetable kingdom are practically infinite.

Yet we learn from the science of chemistry, which deals with the composition of all these objects, that only some seventy odd **elementary** or simple substances exist, and that by combinations of these few elements the countless varieties of things we see are produced.

The science of chemistry then has for its object the study of the properties of the elements and the discovery of the laws ruling their combinations one with another. Beginning with **elements**, we look upon these only as distinct kinds of matter; since they cannot be resolved into simpler kinds of substances, nor formed from the union of other substances, nor changed by any means into each other. The belief in any possible transmutation of elements finds no place in modern chemistry.

But chemistry is experimental, and attempts are constantly made to break up or **decompose** these elements; should such attempts succeed in any instance, the substance so decomposed would of course cease to be looked upon as elementary; but so long as the attempts fail we must regard our present stock of elements as ultimate simple bodies.

We have no power to increase or diminish in any degree the quantity of matter: as regards chemical elements, we can neither create nor destroy. All matter is subject to attraction by other matter, and the attraction of the earth or **gravitation** is measured or expressed as weight. Now in chemical change a given mass or weight of an element may pass into combinations of any kind, but the mass or weight of the element cannot be changed. We express this fact by saying that Matter is indestructible.

When elements unite, the bodies produced are called **compounds**, and the change is called a **combination** of the elements. A compound may be formed of two elements or a greater number; thus water contains two elements (oxygen and hydrogen), sugar contains three (oxygen, hydrogen, and carbon), and four, five, or more elements may enter into complex substances.

To illustrate what is intended by **combination**, let two elements, iron and sulphur, be brought together by rubbing iron filings with flowers of sulphur in a mortar. In this way we produce a **mixture** in which the two substances are each present still in the elementary form but not combined with each other. Such a mixture is easily separated into the materials from which it was made;

the iron particles can be picked out by a magnet, or the two can be separated by washing, or one may be dissolved by a solvent, leaving the other. If also we examine such a mixture under a microscope, the particles of iron and of sulphur are seen to be still there. But if the mixture is heated strongly in a glass tube, combination will take place, and a chemical compound, viz. sulphide of iron, will be produced. No longer is it possible to separate the elements by any such simple methods as sifting or washing, nor can iron or sulphur be recognised under the microscope. In the compound we have a substance apparently of a new kind formed by chemical action.

Another example of a mixture which is of a purely mechanical kind is gunpowder, produced by intimately mixing sulphur and charcoal with nitre. To the eye it is a uniform mass of similar grains, but we can easily separate the three ingredients as follows :—Boil a little powder in water and pass the liquid through a filter; the nitre dissolves in the water and passes through the paper, leaving behind only a mixture of sulphur and charcoal.

These, after being dried, are separated by boiling with carbon di-sulphide which dissolves only the sulphur, and filtering the solution. The charcoal is left on the filter paper, while the sulphur passes through and can be obtained by leaving the solution exposed to the air; in a short time the volatile liquid disappears, leaving crystals of sulphur. In a similar way the solution of nitre, by gentle heating, may be evaporated until crystals are obtained. Thus by simply dissolving the constituent parts in a suitable way we separate our gunpowder into the substances from which it was made.

But let another portion of powder be set on fire; instantly a violent chemical action takes place; the charcoal and sulphur are burnt, and their separation from the gases formed would be a very difficult and tedious operation, and impossible by the simple means used with the mixture.

When a chemical union takes place we usually notice a change of temperature. Heat is in most cases given out (less frequently absorbed), so that the acting bodies become warmer. For instance, let dilute sulphuric acid be poured upon zinc in a flask; a gas (hydrogen) is seen to escape with effervescence, and the metal gradually dissolves in the acid, which becomes warm as the action goes on; a thermometer shows the rise of temperature, which can also be felt when the hand is placed on the flask. The burning of gas, a candle, a fire, etc. are chemical actions of great intensity, producing, besides sensible heat, light also.

The amount of energy in the form of heat produced by chemical actions has been measured by finding how much a given quantity of water is warmed. A **thermal unit** is the heat required to raise the temperature of **a gramme of water from $0°$C. to $1°$C.** If hydrogen gas is burnt in a platinum vessel under water, the heating power of its union with oxygen may easily be found, and measured by the amount the water is warmed. This value is about 34,000; that is, a gramme of hydrogen (one unit weight) in burning is capable of heating 34,000 grammes (or unit weights) of water one degree [1].

Besides the change of temperature produced by chemical action, there are usually changes in the physical properties of the elements used. A compound is not in its characters like the substances out of which it is formed, but as a rule is entirely different. Many examples of

---

[1] The thermal unit or *caloric* is expressed by the letter C., and this value is written as 34,000 C.

this will be noticed later, but we may mention one instance—the differences between bright iron and the rust which forms when it unites with oxygen from the air.

Laws of Chemical Combination.

Thus far we have only considered certain physical differences between **mixtures** and **compounds**; but a most important distinction is found when we consider the quantities of the substances which may be used in either case. For although we can mix elements in all proportions, they will only combine in fixed and **definite proportions**. Thus if we rub together sulphur and mercury in a mortar, we can vary the quantities at will; but if the mixture be heated, the elements only unite in definite quantities, and any excess of either is left and cannot enter into the compound. And the first law of chemical combination is that it always takes place in definite proportions. It is proved by analysis that a compound, whatever be its origin or mode of preparation, is always composed of the same proportions of elements and has always the same composition.

Thus common salt contains (when pure) 60·61 per cent. of sodium and 39·39 per cent. of chlorine, and no variation from these proportions is ever known. And a similar constancy of composition is found in every definite compound.

Combination in Multiple Proportions.

Although the proportions in which elements unite are definite, it is possible to have compounds of two or more elements in more than one ratio: thus carbon and oxygen unite and form two distinct oxides, expressed in symbols as CO, CO_2, and the second contains twice as much oxygen as the first, the quantity of carbon being the same in both; or mercury and chlorine unite to form two distinct chlorides, written $HgCl_2$ and Hg_2Cl_2, in which the mercury is doubled while the chlorine is the same in each. And numerous other examples are met with. But in all such cases the proportions of the variable element bear a simple *multiple* relation to each other.

COMBINATION IN RECIPROCAL PROPORTIONS.

We find also that the proportions in which elements combine are reciprocal towards one another, and can be exchanged in compounds. Thus suppose we know that any three elements combine: if the proportion of the first pair is expressed by the ratio $a : b$ and the second by $b : c$, we shall find that the first and third unite in the ratio $a : c$ [or in a simple multiple of this ratio].

We may take as an example the elements mercury, oxygen, and chlorine, and express the ratios in which they combine thus :—

$$Hg : O : Cl_2 = 200 : 16 : 71.$$

Then

    I. Mercuric oxide HgO = 200 : 16.
    II. Mercuric chloride $HgCl_2$ = 200 : 71.
    III. Chlorine oxide OCl_2 = 16 : 71.

When we know the first two ratios we can infer that if the third compound is formed, oxygen and chlorine will unite in the ratio of 16 : 71 (or simple multiples of these numbers).

By analysis we can discover what elements are present in any substance, and further we can find out how much of each is present. For the first purpose analysis is said to be **qualitative**, and for the second **quantitative**.

Since the elements combine only in fixed and definite proportions and these quantities stand in reciprocal relations, we can express how much of one element combines with another element by numbers, and these numbers are the **combining weights** of the elements. They express only the relations of the weights which can combine with one another; the numbers thus determined are purely experimental, and do not depend upon any theory of atoms. Sometimes the expression **equivalent weight** is used for the same purpose, to express either the weights of elements which combine together, or the weights which replace each other in reciprocal combinations. In the example just given we may say that 16 parts of oxygen are **equivalent** to 71 parts of chlorine, since they each combine with 200 parts of mercury.

Atoms and Molecules.

In order however to explain the laws of chemical combination, John Dalton in 1808 proposed the **Atomic Theory**, which with slight change is the explanation now generally accepted. But it must be kept in mind that this 'hypothesis' for chemical changes is not a matter which can be verified and proved by experiment as the laws of chemical combination are, but is a step beyond our experimental methods and an attempt to explain the reasons for our experimental results.

It is supposed then that matter cannot be divided to an infinite extent, but that it consists ultimately of indivisible and indestructible particles which are primitive **atoms**. All atoms of any element are alike, and for each element the atom has a definite mass and weight. When chemical union takes place, we suppose the atoms unite by each one joining itself to one other, or to two others or three others as the case may be; so that every chemical compound has a constant composition, since the proportions in which the atoms combine are constant. Thus, if one particle of water contains two atoms of hydrogen and one atom of oxygen, every other particle is similarly composed, and however great or small the number of particles, since the composition of each is constant, so is the composition of the whole. We have used the term particle to express the compound formed by atomic combination of the elements, but a more convenient and common term is **molecule**. We mean by a molecule of water the unit of water which is produced by the union of the atoms, so that the molecule is the primitive particle of water; which if broken up would divide only into the atoms of the elements composing it.

Avogadro's Hypothesis.

Next in importance to the Atomic Theory as a help to explaining chemical action is the hypothesis of Avogadro put forth about the year 1811. He supposes that all gases under similar conditions of pressure and temperature are composed of particles (or molecules) in motion and also that the number of molecules in a given volume is equal for every gas, under similar conditions. Thus a litre flask, whether filled with hydrogen or oxygen, or ammonia or steam, etc., contains the same number of molecules under similar conditions of temperature and pressure. It is a fact proved by abundance of experiments that gases combine with each other in simple relations of volume, that is, in equal volumes or one volume to two, etc., and this fact is easily understood when we know that in equal volumes of these are the same number of molecules [1].

---

[1] This number is estimated to be 10^{18} molecules in 1 cc. of gas.

Atoms and Molecules.

Some important consequences follow from this hypothesis. Suppose we take equal volumes of hydrogen and of chlorine, and mix them and then expose the mixture to light; the compound hydrogen chloride is formed, and no alteration of volume takes place. We can represent this by a diagram:—

We see that in the **compound** in a given space we shall have a certain number of molecules made up of dissimilar atoms (H, Cl) while in the same space when filled with an **elementary** gas there are (by hypothesis) the same number of molecules which are made up of similar atoms (H, H) and (Cl, Cl).

From this reasoning we arrive at the conclusion that a molecule of hydrogen is H_2 and a molecule of chlorine is Cl_2: and we are in a position to determine the weight of the molecule of any other element or compound in the gaseous form by weighing some defined volume and comparing the weight with that of the same volume of hydrogen at standard temperature and pressure. The following examples will illustrate this, but many others will be given later.

It is found by experiment that 22·4 litres of hydrogen (t=0° C., Bar. 760 mm.) weigh 2 grams: we will use this volume for comparison.

Atomic Weights.

Symbol.		I. Experimental Weight of 22·4 litres.	II. Theoretical Molecular Weight.	III. Experimental Vapour Density. H = 1.
		grains.		
H_2	Hydrogen	2	2	1
O_2	Oxygen	32	32	16
N_2	Nitrogen	28	28	14
Cl_2	Chlorine	71	71	35.5
HCl	Hydrogen Chloride	36.5	36.5	18.25
H_2O	Steam	18	18	9
NH_3	Ammonia	17	17	8.5
H_4C	Marsh Gas	16	16	8.0

Column III shows the vapour density in each case as compared with hydrogen taken as unity. The vapour density is one half of the molecular weight, and is calculated directly from the experimental results in column I: it gives the ratio of the weights of equal volumes of the gases compared with hydrogen as unity.

Atomic Weights.

The numbers given as the atomic weights of the elements are intended to express the relative weights of the atoms; taking the weight of an atom of hydrogen as unity. They obviously cannot be found directly, but are determined usually from the following data :—

(1) The equivalent or combining weight of the element is found by analysis or synthesis of its compounds.

(2) The molecular weight is calculated from the vapour density of the element or some compound.

(3) The specific heat of the element is determined, if possible. The application of these data to the cases of the principal elements and their compounds are separately discussed hereafter.

The following tables give the elements with their

Atomic Weights.

symbols and atomic weights as far as they are at present known; but we have only to deal with those of frequent occurrence. As a starting-point we take the elements composing water and air, beginning with hydrogen.

Hydrogen	H	1
Lithium	Li	7
Beryllium	Be	9
Boron	B	11
Carbon	C	12
Nitrogen	N	14
Oxygen	O	16
Fluorine	F	19
Sodium	Na	23
Magnesium	Mg	24
Aluminium	Al	27
Silicon	Si	28
Phosphorus	P	31
Sulphur	S	32
Chlorine	Cl	35.5
Potassium	K	39
Calcium	Ca	40
Scandium	Sc	44
Titanium	Ti	48
Vanadium	V	51
Chromium	Cr	52
Manganese	Mn	55
Iron	Fe	56
Nickel	Ni	58.6
Cobalt	Co	59
Copper	Cu	63
Zinc	Zn	65
Gallium	Ga	69
Germanium	Ge	70
Arsenic	As	75
Selenium	Se	79
Bromine	Br	80
Rubidium	Rb	85
Strontium	Sr	87
Yttrium	Yt	89
Zirconium	Zr	90
Niobium	Nb	94
Molybdenum	Mo	96
Rhodium	Rh	104
Ruthenium	Ru	104.5
Palladium	Pd	106
Silver	Ag	108
Cadmium	Cd	112
Indium	In	114
Tin	Sn	118
Antimony	Sb	120
Tellurium	Te	126
Iodine	I	127
Caesium	Cs	133
Barium	Ba	137
Lanthanum	La	139
Cerium	Ce	141
Didymium	Di	144
Samarium	Sm	150
Holmium	Ho	160
Erbium	Er	166
Ytterbium	Yb	173
Tantalum	Ta	182
Tungsten	W	184
Iridium	Ir	192.5
Osmium	Os	193
Platinum	Pt	194.5
Gold	Au	196
Mercury	Hg	200
Thallium	Tl	204
Lead	Pb	207
Bismuth	Bi	209
Thorium	Th	232
Uranium	U	240

Symbols are used to represent the elements, and a combination is expressed by writing symbols side by side; thus, HgO means a compound of mercury and oxygen. The plus sign is used for simple admixture; Hg + S signifies a mixture of mercury and sulphur.

The **formula** of a compound is intended to show its composition, and it expresses the elements present in a compound and also the **quantity** of each; every symbol for an element being used to indicate the atomic weight of that element. Thus the formula for water is H_2O, and signifies that water is a compound of hydrogen and oxygen, and taking the atomic weight of hydrogen as 1 and that of oxygen as 16, we know from the formula that water contains 2 parts by weight of hydrogen and 16 parts by weight of oxygen. Again, potassium chlorate is represented by the formula $KClO_3$, and by using the atomic weights given in the table, we get

$$
\begin{aligned}
\text{Potassium} &= K = 39 \cdot 0 \text{ parts by weight.} \\
\text{Chlorine} &= Cl = 35.5 \quad ,, \quad ,, \\
\text{Oxygen} &= O_3 = 48 \quad ,, \quad ,, \\
\text{Potassium Chlorate} &= KClO_3 = 122.5 \quad ,, \quad ,,
\end{aligned}
$$

From these numbers it is obviously easy to find the **composition per cent.** of the substance; thus

$$
\begin{aligned}
\text{Potassium} \quad & \frac{39}{122.5} \times 100 = 31.84 \text{ per cent.} \\
\text{Chlorine} \quad & \frac{35.5}{122.5} \times 100 = 28.98 \quad ,, \\
\text{Oxygen} \quad & \frac{48}{122.5} \times 100 = 39.18 \quad ,, \\
& \hspace{4.5cm} \overline{100.00}
\end{aligned}
$$

Chemical changes are usually represented by **equations**, and the substances on the left represent the beginning and those on the right the end of the reaction.

Thus the simple **combination** of hydrogen with oxygen to form water is represented by the equation

$$H_2 + O = H_2O,$$

and the **decomposition** of water into its elements is written
$$H_2O = H_2 + O;$$
or we may have reactions in which two compounds take part, such as
$$AgNO_3 + NaCl = AgCl + NaNO_3.$$

Silver nitrate with sodium chloride forms silver chloride and sodium nitrate; this is an example of **double decomposition**.

Since the equation represents the *quantitative* changes as well as the *qualitative*, we can easily arrive at the quantities of the reacting bodies and also of the products. Thus:—

Ag	= 108	Na	= 23	Ag	108	Na	23
N	= 14	Cl	= 35.5	Cl	35.5	N	14
O_3	= 48					O_3	48
$AgNO_3$	= 170	NaCl	= 58.5		143.5		85

Or 170 parts by weight of silver nitrate, with 58.5 parts of sodium chloride, produce 143.5 parts of silver chloride and 85 parts of sodium nitrate; and the total weights on opposite sides are identical,
$$170 + 58.5 = 143.5 + 85.$$

CHAPTER II.

**Hydrogen. H. Atomic weight = 1. Density = 1.
Molecular weight = 2 = H_2.**

HYDROGEN in the free gaseous state is only found in nature in small quantity issuing from earth crevices in volcanic districts, or near petroleum wells.

It exists in combination everywhere; as a constituent of water, of all plants and animals, and in numerous minerals, abundantly in coal, petroleum, bitumen, etc., and to a lesser degree in rocks. The element may be separated from any of its compounds, but is usually obtained from water or dilute acids. For laboratory purposes it is most convenient to prepare the gas by acting upon dilute sulphuric acid with zinc.

A flask is provided with a well-fitting cork bored with two holes; one is fitted with a bent glass tube to carry off the gas, the other admits a thistle funnel which passes nearly to the bottom of the flask, and serves for a safety tube. Granulated zinc is placed in the flask, and, the apparatus being fitted together, dilute sulphuric acid is poured down the thistle funnel; fig. 1. A convenient strength is one part of acid to seven of water; strong acid must not be used. At once the acid begins to dissolve the metal and a gas comes off briskly with effervescence. As the action proceeds, the liquid will be found to get warm, and the gas will come off more quickly in consequence of the rise in temperature. At first the hydrogen is mixed with air from the flask, and the first portions of the gas should not be collected; after a few minutes the gas can be collected over water in the pneumatic trough, and several jars filled for experiments.

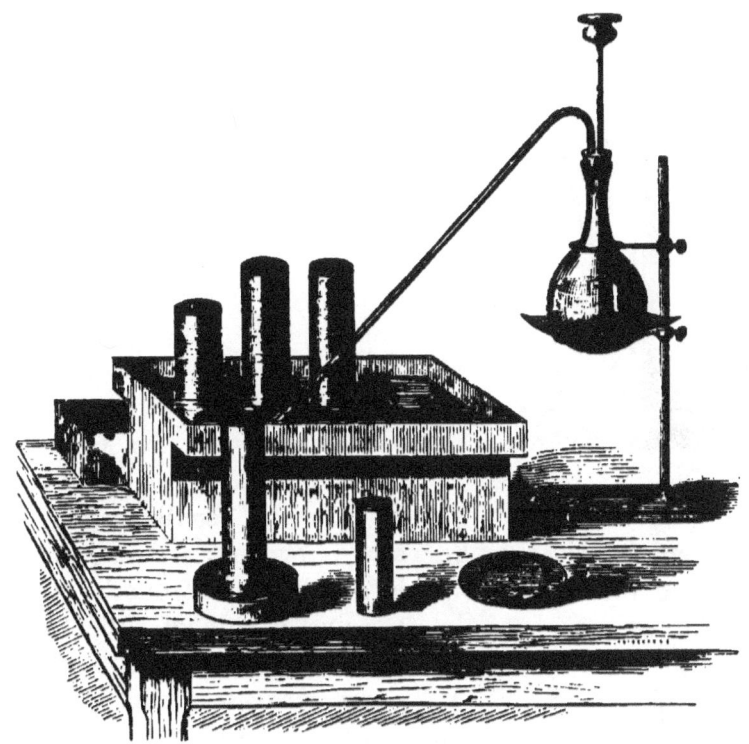

Fig. 1.

We can represent the chemical action in symbols by an equation thus:—

$$Zn + H_2SO_4 = ZnSO_4 + H_2.$$

Zinc and sulphuric acid become zinc sulphate and hydrogen. The metal zinc displaces hydrogen from the sulphuric acid and is converted into zinc sulphate, which dissolves in the water present. If we wish to obtain the zinc sulphate, we pour off the solution and

Decomposition of Water.

filter it (to remove any particles of solid matter or any impurities from the zinc), and gently evaporate the liquid so as to get rid of the water: finally, white crystals of zinc sulphate will be obtained.

This method of preparing hydrogen may be varied by using hydrochloric acid in the place of sulphuric, or we may use the metal iron instead of zinc with either acid.

Hydrogen can be obtained from water in several ways.

I. Water is decomposed by electricity. When an electric current from three or more cells of a battery (such as Grove's) is passed between two platinum plates in water, containing some ($\frac{1}{10}$) sulphuric acid, the water is decomposed; hydrogen being given off from one plate and oxygen from the other.

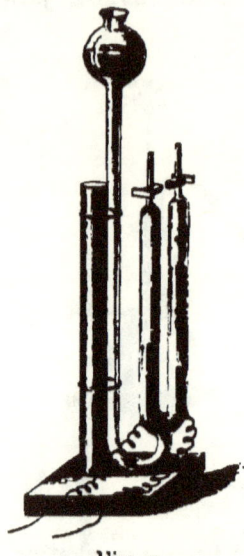

Fig. 2.

By using a pair of tubes over the platinum plates we can collect the gases separately and examine their properties (fig. 2).

II. Hydrogen is liberated from water, even when cold, by certain metals.

Fig. 3.

A piece of potassium thrown on water floats about and acts so energetically that the liberated hydrogen takes fire. Sodium behaves in a similar manner, but is less active. If we roll a piece of paper tightly round a small stick of sodium, fastening it so as to leave one end open, and hold it under an inverted tube filled with water, the hydrogen given off is caught and can be examined (fig. 3).

The change is thus represented :—

$$2H_2O + Na_2 = 2NaOH + H_2.$$

Water and sodium yield sodium hydrate and hydrogen.

III. Hydrogen is set free from water by iron at a red heat.

For showing this, an arrangement shown in the figure 4 may be used.

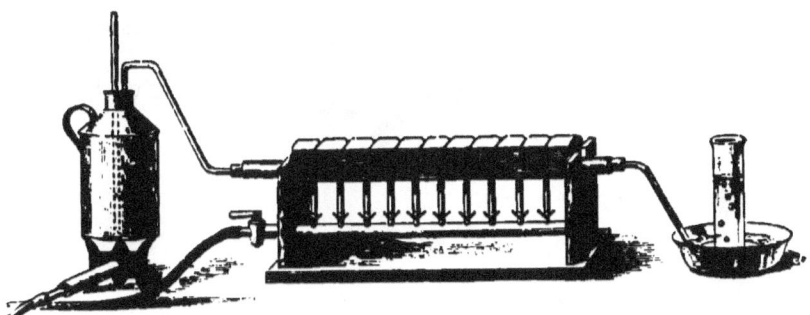

Fig. 4.

A length of iron gas-pipe is filled with iron nails and placed in a gas-furnace by which it can be made red hot. Steam is obtained from a tin can heated by a separate burner and passed through the iron tube, the hydrogen being collected over water in the trough. The tin vessel should be provided with a glass tube open at each end and reaching nearly to the bottom to act as a safety valve.

The chemical change in this case is shown by the equation

$$3Fe + 4H_2O = Fe_3O_4 + 4H_2.$$

Water and iron produce iron oxide and hydrogen.

18 *Properties of Hydrogen.*

Note. If the nails are already coated with oxide they should be heated to redness in the tube and reduced by hydrogen gas or common coal gas.

Properties and Characters.

Hydrogen gas thus obtained is colourless and invisible; it burns in air with a pale flame forming water, and mixed with air or oxygen makes a highly explosive

Fig. 5.

mixture. It is the lightest of all elements, and is for this reason adopted as a standard of density. Taking hydrogen as unity, the density of air is about $14\frac{1}{2}$. It was at one time thought to be a permanent gas, but by great pressure and a great reduction of temperature it has been obtained as a liquid. All non-metallic elements and many metals form compounds with hydrogen.

The flame of burning hydrogen is extremely hot, and if oxygen gas is driven into it the heat becomes so intense that few substances resist its melting power.

Hydrogen.

Silver, gold, and even platinum melt and boil in the flame. Lime does not melt, but glows vividly; and the intense brilliant light so produced (**lime light**) is often used for illuminating purposes.

The following experiments may be performed with the gas to show its properties :—

1. A toy balloon or soap-bubbles filled with hydrogen rise rapidly in air.

2. The gas escapes slowly from a jar held mouth downwards; instantly if held mouth upwards: or hydrogen can be poured from one jar *upwards* into another (fig. 5).

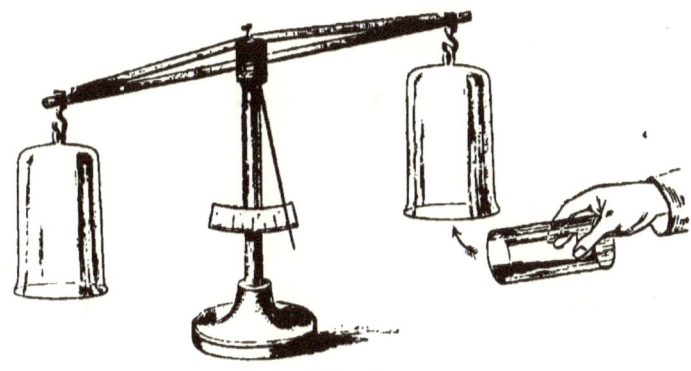

Fig. 6.

3. Let two beakers be suspended on a balance mouth downwards; if a jar of hydrogen be poured *upwards* into one of them, the arm of the balance will rise on that side (fig. 6).

4. The combustible character of the gas is shown by lighting a jar of the gas; if pure, it burns quietly without explosion.

5. Pass a lighted taper into a jar of hydrogen, held mouth down; the gas kindles, but the taper will not burn in the gas, but will light again as it is withdrawn.

6. A small jet of oxygen or air brought into a jar of burning hydrogen will inflame and burn within the hydrogen gas (fig. 7).

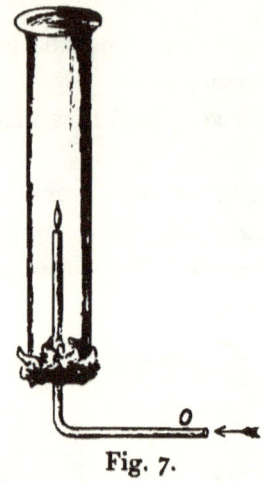

Fig. 7.

Calculation of the Weight and Volume of Hydrogen.

The equation for the formation of hydrogen given above is

$$Zn + H_2SO_4 = ZnSO_4 + H_2.$$
$$65 + 2 + 32 + 64 = 65 + 32 + 64 + 2;$$

and by filling in the atomic weights of the elements in the equation we see that 65 parts of zinc with 98 of sulphuric acid yield 161 of zinc sulphate and 2 parts of hydrogen. We are able therefore to calculate from any weight of zinc what weight of hydrogen will be given off.

And the volume of 2 grammes of hydrogen is 22·4 litres, or one litre of hydrogen weighs ·0896 grammes.

Note. The heat produced by burning one gramme of hydrogen is 34,000 c. See page 4.

CHAPTER III.

**Oxygen. O. Atomic weight, 16. Density, 6.
Molecular weight** $= 32 = O_2$.

OXYGEN is the most abundant of the elements; being in actual weight nearly one-half of the total quantity of the outside portions of the earth of which the composition is known to us. In the free uncombined state it forms about one-fifth of the atmosphere; in the combined form it is eight-ninths of the mass of the water of the globe; it is also present as the chief constituent in every geological formation except coal. In granite, slate, clay, limestone and other rocks, the quantity of oxygen reaches nearly to 50 per cent.

In the atmosphere, oxygen is mixed with four times its volume of nitrogen, and the isolation of this element and its separation from air by Priestley in 1774 proved that air is not a simple and elementary body, as was the general belief up to that time.

We may obtain oxygen from air by Priestley's method; by heating mercury to boiling in air it becomes covered with red scales (of oxide), and these red scales being more strongly heated will break up into metallic mercury and oxygen (fig. 8).

$$HgO = Hg + O.$$
Mercury Oxide = Mercury and Oxygen.

The substance most used as a source of oxygen is potassium chlorate. A small quantity of this salt well dried, heated in a hard-glass test-tube, will furnish a supply of the gas, which can be collected over water.

The salt when heated melts, and at first the gas comes off briskly; after a time the salt appears less fusible, and finally solidifies unless a very strong heat be used—

$$KClO_3 = KCl + O_3.$$
Potassium chlorate = potassium chloride and oxygen.

[Potassium perchlorate $KClO_4$ is formed and decomposed during the operation: see page 129.]

When potassium chlorate is mixed with manganese

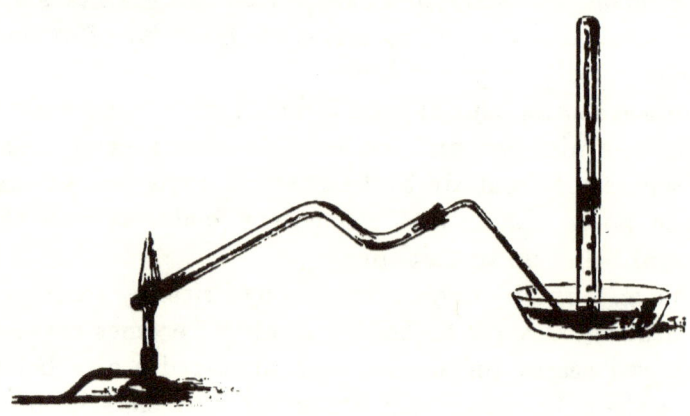

Fig. 8.

di-oxide—MnO_2—the oxygen gas comes off very quickly and at a much lower temperature, and in laboratories such a mixture is as a general rule employed. If considerable quantities of gas are wanted, a retort of iron or copper is useful for heating the mixture. Oxygen thus made is often contaminated with chlorine; it should be purified by passing through a wash-bottle containing solution of soda (fig. 9).

A Pepys gasholder represented in the figure is a convenient receptacle and store vessel. Oxygen gas compressed into strong wrought iron cylinders is now an article of commerce.

Oxygen may be obtained from numerous other substances by simple heating; oxide of silver, like oxide of mercury, is broken up; peroxide of manganese loses one-third of its oxygen at a red heat ($3\,MnO_2 - O_2 = Mn_3\,O_4$). Many metallic salts decompose by heat, yielding oxygen with other products (sulphates, nitrates, chromates, etc.).

It has already been noticed that water yields oxygen

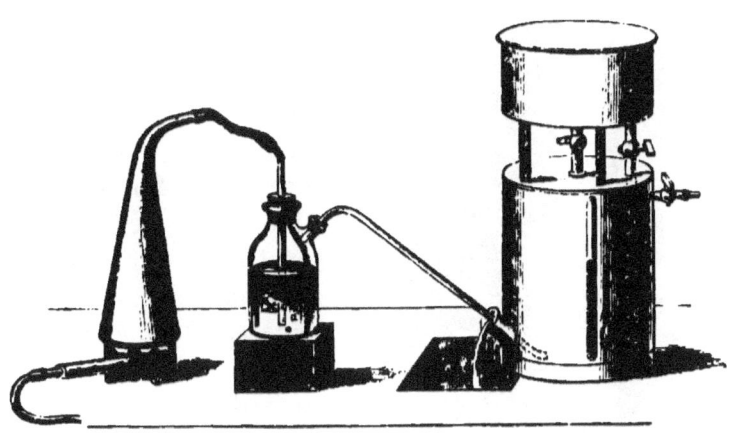

Fig. 9.

when the electric current is passed through it: by separating the battery poles with a porous earthen cell the gases can be collected separately.

Properties. Oxygen is a gas without colour, taste, or odour; it may be liquefied by great pressure combined with a very low temperature. The liquefied oxygen boils at $-182°$C. (pressure = 760 mm.).

The density of oxygen gas is 16 (hydrogen = 1), and the atomic weight of the element is 15·96, or approximately also 16. It combines with all other elements, except only fluorine: the compounds or *oxides* thus formed being acid-forming oxides or basic oxides, etc., as we shall see later. Water dissolves a little (about 4 per cent.) of the gas, and oxygen is therefore taken up by all water freely exposed to the atmosphere. Fishes and other aquatic animals and plants use the dissolved oxygen for respiration.

Fig. 10.

The oxygen in our atmosphere is necessary to the respiration of all air-breathing animals and plants, and is also necessary for combustion. All bodies combustible in air will burn in oxygen with increased brilliancy and rapidity. We may illustrate this by the following experiments:—

(1) Carbon, (2) sulphur, (3) phosphorus can be burned in oxygen; a deflagrating spoon is convenient for the purpose (fig. 10).

The actions are—

$$C + O_2 = CO_2. \quad S + O_2 = SO_2. \quad P_2 + O_5 = P_2O_5.$$

(4) A piece of steel wire or watch-spring may be burnt in a jar of oxygen. In order to make the metal sufficiently hot at the end to start the experiment, a little cotton thread should be twisted round the tip and

fastened off and then dipped into a little melted sulphur. This being ignited and brought into a jar of oxygen the steel wire will burn vividly (fig. 11).

Fig. 11.

CALCULATION OF THE WEIGHT AND VOLUME OF OXYGEN.

From the equation $2HgO = 2Hg + O_2$,

$$Hg = 200, \ O = 16,$$

we see that 216 grammes of mercuric oxide are broken up into 200 g. of mercury and 16 g. of oxygen; or 432 g. of oxide give 32 g. of oxygen.

Since

```
      1 gramme  of hydrogen = 11.2 litres
  ∴ 16 grammes  of oxygen   = 11.2   ,,
    and 32   ,, of oxygen   = 22.4   ,,
```

From the equation $KClO_3 = KCl + O_3$, by assigning atomic weights to each element, we get

Potassium	. .	39.0	Potassium	. .	39.0
Chlorine	. . .	35.5	Chlorine	. . .	35.5
Oxygen	. . .	48.0			

Potassium chlorate 122.5 yields Potassium chloride 74.5

and 48 parts of oxygen are given off; and 48 grammes of oxygen will measure $3 \times 11.2 = 33.6$ litres.

Ozone. The action of electricity upon oxygen produces a remarkable modification or *allotropic* form which has been named ozone. In the neighbourhood of an electric machine at work a peculiar smell may be noticed; the oxidation of phosphorus and of some organic substances produces a similar effect. And although ordinary oxygen has no odour, yet the electrified gas and the oxygen liberated from water by the electric current possess the peculiar odour mentioned.

If pure dry oxygen is acted upon by the electric discharge it contracts slightly in bulk, about one-twelfth as a maximum, but only a portion of the gas can be converted into ozone. A convenient form of ozone tube is shown in the figure 12. A thin glass tube like a test tube is sealed within a similar slightly larger tube and inlet and exit tubes are sealed into the outer tube. The inner tube contains brine (as a conducting liquid, in which also the apparatus is immersed). On connecting the inner and outer liquids with wires from a Ruhmkorff coil, as though they were the coatings of a Leyden jar, they are separated by the two glass walls of the tubes with a layer of gas between, which becomes thus strongly electrified. A stream of oxygen, carbon dioxide, or carbon monoxide passing through the apparatus becomes strongly charged with ozone.

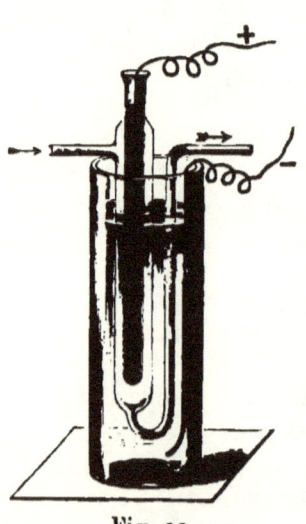

Fig. 12.

Ozone is a very active form of oxygen; it liberates iodine from solution of potassium iodide,

$$2KI + H_2O + O_3 = 2KOH + I_2 + O_2,$$

and oxidises mercury and silver. It is destroyed by heat expanding to its original bulk and becoming converted into ordinary oxygen. Oil of turpentine absorbs it entirely. Ozone cannot be obtained unmixed with common oxygen, but from the relation between its oxidising power (in setting free iodine) and the volume of gas absorbed by oil of turpentine the molecule appears to be represented by O_3, the density being $\frac{48}{2} = 24$.

CHAPTER IV.

**Water; H_2O. Molecular Weight = 18.
Density of Vapour = 9.**

WATER exists in nature as *solid* ice or snow; in the *liquid* form in oceans, lakes, streams; and as a *vapour* in the atmosphere. As a true vapour it is invisible, but condensing becomes visible as mist, fog, cloud, rain, or dew. In addition, water is contained within the earth in enormous quantity; and the underground water is a great store which we use when it issues in springs, or which can be reached in some cases by wells and borings.

Water is not an element, since it can be formed by the union of hydrogen and oxygen; and also can be broken up into these substances. We can resolve water into its constituent elements by the electric current. When two platinum plates are immersed in water (containing one-tenth part of sulphuric acid to make it conduct) and the plates are joined to the poles of a batttery a brisk evolution of gas takes place (fig. 13). If the mixture of hydrogen and oxygen is collected and fired an explosion takes place, and water is reproduced.

We may also decompose steam by a series of electric sparks from an induction coil, but no large quantity of the gases can be obtained. At a very high temperature —above 1000° C.—steam is broken up into the gases of which it is composed.

The formation of water from its constituent elements,

Synthesis of Water.   29

or the *synthesis* of water, may be effected in a variety of ways.

(1) Hydrogen burns in the air, producing water. This is easily shown by burning a jet of hydrogen gas beneath a *large* glass shade and collecting the condensed drops of water, which soon trickle down the glass. We can prove similarly that water is formed when coal-gas or a candle or any similar combustible containing hydrogen is burnt.

(2) Oxygen gas will burn in hydrogen, forming water.

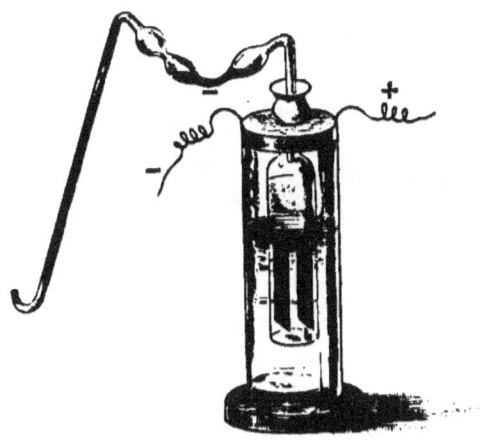

Fig. 13.

Let a tall narrow jar full of hydrogen be held mouth down and the gas ignited; on bringing a jet of oxygen from a gas-holder into the flame and up into the jar, the oxygen gas will take fire and burn within the jar (see fig. 7).

(3) Hydrogen reduces metallic oxides, forming water.

A little oxide of copper is placed in a hard-glass tube (d) connected with a hydrogen generator (a) (fig. 14). The oxide is not reduced so long as it is cold, but a gentle heat being applied while the current of hydrogen passes brings about the reduction: the copper parts with its oxygen (being reduced to the metallic state), and water is formed, which condenses in a U-tube (*e*) attached to the apparatus. Oxides of iron and lead can be reduced in a similar way.

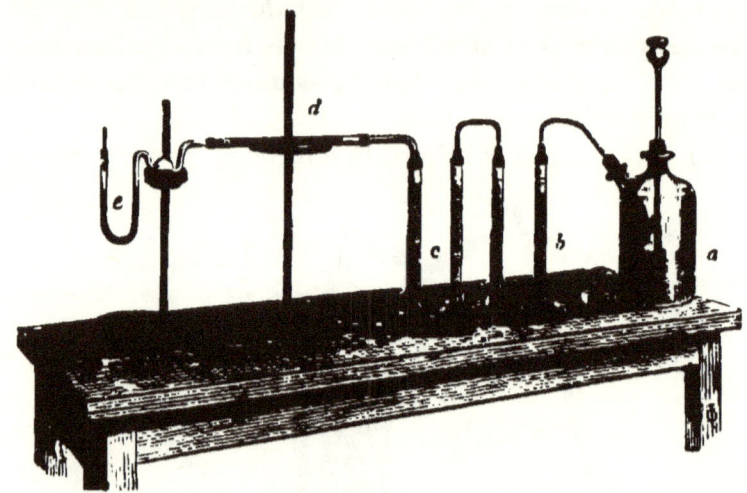

Fig. 14.

Composition of Water by Weight.

The proportion in which hydrogen and oxygen combine has been determined with the greatest care and accuracy, since this ratio enters into the calculation of the atomic weights of a large number of elements. The process is found upon the reduction of copper oxide when heated in hydrogen.

The apparatus used consists of a series of parts which are represented in the figure 15 :—

1. An apparatus for generating hydrogen.
2. Tubes for purifying and drying the gas.
3. A tube to prove that the gas is perfectly dry.

Weights of Components. 31

4. Bulb to contain copper oxide.
5. Tubes to catch and retain all the water formed.
6. A tube to prove that the gas is dry and all the water is retained in 5.

Before the experiment begins the bulb containing the perfectly dry copper oxide must be weighed, and the tubes with the drying substances must be weighed. The weights also of the proof-tubes are noted.

Now when the gas is turned on and the apparatus filled with hydrogen the bulb is heated; the reduction of the oxide and the formation of water begin. After a

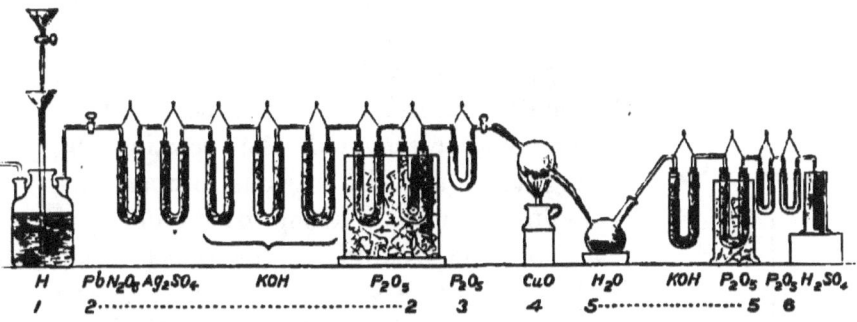

Fig. 15.

time the heating is stopped and the apparatus cooled. The bulb being weighed will be found lighter by reason of OXYGEN lost; the tubes 5–5 will be heavier by reason of WATER gained. The difference between the weight of water and of oxygen gives the HYDROGEN.

The small tubes being re-weighed should remain unchanged, thus proving the perfect action of the drying substances.

Some actual experiments performed by Dumas, every precaution against error being taken, gave—

Oxygen consumed = 840.16 grammes;
Water produced = 945.44 ,,

From which by calculation we deduce

$$\begin{array}{rl} \text{Oxygen} =& 88.86 \\ \text{Hydrogen} =& 11.14 \\ \hline \text{Water} =& 100.00 \end{array}$$

Or supposing the amount of hydrogen = 1, the oxygen combined with it = 8, or two parts by weight of hydrogen unite with 15·96 parts of oxygen, or nearly 16.

Composition of Water by Volume.

If 16 grammes of oxygen unite with 2 grammes of hydrogen, the volume of oxygen is 11·2 litres, and the hydrogen is 22·4 litres, or twice as large. This gives us indirectly the composition by volume; but the fact can also be determined by direct experiment.

Cavendish discovered the relation by mixing the gases in various proportions, exploding them in a strong glass globe, and measuring the residue; he found that when the hydrogen was twice the volume of the oxygen no gas was left after the explosion.

It is however easier to explode a mixture and measure the residue.

The eudiometer, more generally employed for exact gas analysis, is a long glass tube (500–600 mm.) closed at one end, with platinum wires sealed into the glass near the top and graduated into divisions 1 mm. apart (fig. 16). The capacity of the tube is found by direct measurement for several points, and a table is constructed giving the volume for every division.

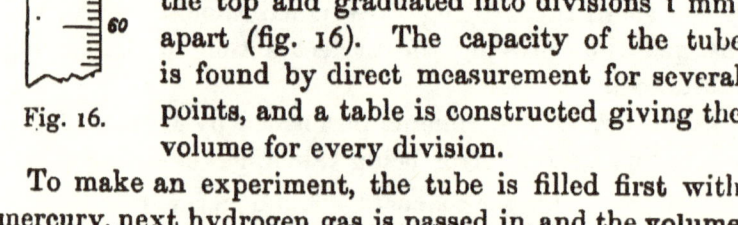

Fig. 16.

To make an experiment, the tube is filled first with mercury, next hydrogen gas is passed in, and the **volume**

Composition of Water. 33

occupied and the **temperature** observed (fig. 17). The **pressure** of the gas needs to be found as follows :—A reading of the barometer gives the external pressure (b); the height of the internal column of mercury is given by readings on the engraved scale of the tube (a), and the actual pressure within is the difference of the heights of these two columns, or ($b-a$). The volume, temperature and pressure of the hydrogen being found, some pure oxygen is passed into the tube and a similar set of readings made. The mixture is now exploded, and

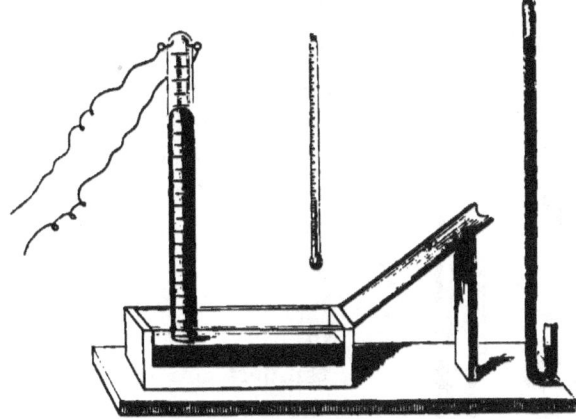

Fig. 17.

fresh readings of the volume, temperature and pressure obtained for the residue of (oxygen) gas.

Since these volumes are found, for unlike pressures and possibly at different temperatures they must be reduced to standard pressure (e.g. 760 mm.) and temperature (0°C.) and corrected for the moisture present, when the true relations of volume appear. (See Chap. XXVIII.)

We may take such an example as the following :—

Volume of hydrogen reduced to 760 mm. and 0°C. = 166
    ,,   hydrogen + oxygen   ,,   ,,   = 280
    ,,   oxygen left (after explosion)   = 31
    ∴ contraction upon explosion   = 249

Volume of Steam formed.

Obviously then 249 volumes of gas were condensed in the explosion, of which 166 were hydrogen and 83 were oxygen, which are in the ratio of 2 : 1.

One-third part of the contraction gives the oxygen, two-thirds gives the hydrogen.

If the water formed, instead of being cooled, is kept above boiling-point, the steam is found to be equal in volume to the hydrogen used. This is done by enclosing the eudiometer with a jacket which can be filled with hot vapour.

A U-shaped eudiometer (fig. 18) is used for the experiment; it is fitted with platinum wires, and has one limb enclosed within a glass jacket; in this manner by passing the vapour of amyl alcohol boiling at a temperature of 130° C. through the apparatus no water can be condensed.

A mixture of oxygen with hydrogen, obtained by decomposing water by an electric current, is brought into the eudiometer and its volume observed when the temperature is constant at about 130°; the mixture is then exploded, and the steam will be found to occupy two-thirds of the volume of the original gas. Note: the volumes are both measured at atmospheric pressure, the mercury being adjusted so as to be at the same level in the two limbs of the eudiometer.

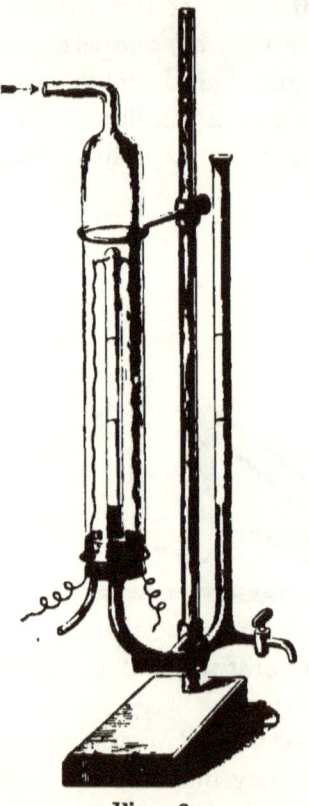

Fig. 18.

We have proved then that water contains two volumes of hydrogen combined with one volume of oxygen, and that two volumes of steam or water vapour are formed when they unite.

The vapour density of steam has also been found by direct experiment to be 9, whence we may infer that the molecular weight of water ($H_2 = 2$) is $18 = H_2O$.

$$\frac{H_2O}{H_2} = \frac{18}{2} = 9 \text{ vapour density of steam.}$$

| H_2 | + | H_2 | + | O_2 | = | H_2O | H_2O |

PROPERTIES OF WATER.

Water is a clear liquid without taste or odour if pure, and with no colour perceptible in small quantities; but viewed through a glass tube with glass ends about two feet in length it appears of a pale blue tint.

When heated, water expands gradually from the temperature of 4°C. until it reaches the boiling point, when it is converted into vapour. At 760 mm. pressure water boils at 100°C., but the temperature at which boiling takes place varies with variations in the pressure. If some water be boiled in a strong globular flask, securely corked while actually boiling, and then taken away from the lamp, boiling will continue within the flask, and by pouring cold water (fig. 19) upon the flask the water boils

Fig. 19.

vigorously: the reason being that the cold water, by condensing some steam, reduces the pressure within the vessel and so lowers the boiling-point. When water is cooled it contracts until the temperature of 4°C. is reached: further cooling causes a slight expansion ($\frac{1}{8000}$) to 0°C., at which point the water solidifies. The freezing of water is marked by a sudden expansion ($\frac{1}{11}$), and this expansion takes place with such force as to burst almost any vessel in which the water is confined. The bursting of pipes in frosts and the splitting of rocks are common examples of the action of the expansive force of freezing water. Since ice expands when it is formed, its density is less than 1 and it floats on water.

$$\begin{array}{rrr} \text{Density of water at} & 0° = & 1. \\ \text{,,} \quad \text{,,} \quad \text{,,} & 100° = & .958 \\ \text{,,} \quad \text{ice} \quad \text{,,} & 0° = & .916 \\ \text{Volume of water ..} & 0° = & 1.000 \\ \text{,,} \quad \text{,,} \quad \text{,,} & 100° = & 1.043 \\ \text{,,} \quad \text{ice} \quad \text{,,} & 0° = & 1.090 \end{array}$$

LATENT HEAT OF WATER.

When a solid changes into a liquid, heat is used up and disappears. The heat required to melt ice is used in doing internal work, or *molecular* work; and as it does not alter the temperature of the ice, cannot be detected by the thermometer. The quantity of heat used up may be measured by finding the quantity of heat lost by warm water when ice is put into it: 100 g. of fragments at 0°C. stirred into 1000 g. of water at 19°C. reduces the temperature to 10°C. when the ice is just melted. Now a unit of heat, or thermal unit, is the heat required to raise one unit of water one degree, so we have as the water was cooled 9°, the loss by water $= 1000 \times 9 = 9000$, the absorption by ice $= 100 x$ (taking $x =$ latent heat of one gramme), and the melted ice is warmed to 10°C. (100 × 10 = 1000);

$$\therefore 100x + 1000 = 9000$$
$$100x = 8000$$
$$x = 80, \text{ or}$$

the latent heat of water is 80; that is, the heat required to melt one gramme (or pound or ton, etc.) of water would raise the temperature of 80 grammes (or pounds or tons, etc.) through one degree.

Latent Heat of Steam.

The conversion of any liquid into the state of vapour requires heat, which being used in doing internal work (molecular work) does not cause a rise of temperature. The heat thus made latent is found by passing steam into water and measuring the rise of temperature, and also weighing the extra water formed.

Thus 626 grammes of water at 0° are warmed to 10° by passing in steam, and the increase in weight of the water is found to be 10 grammes. We have, putting x for latent heat of one gramme of steam—

$$626 \times 10 = 10 \times x + 10 \times 90$$
$$6260 = 10x + 900$$
$$5360 = 10x$$
$$536 = x;$$

or a unit of water requires 536 units of heat to convert it into vapour; a quantity sufficent to raise 536 units of water one degree in temperature.

Solution.

Water is a solvent for nearly all solids, liquids, and gases, but to a varying extent in the case of different substances. Such bodies as sugar, salt, sal-ammoniac are freely soluble in water, while others, such as gold, mercury, metals, glass, etc., are apparently insoluble. But glass, silica, and many substances which may appear insoluble, are really soluble to a minute extent. Hot water dissolves more quickly, and in most cases in larger quantity, than cold; the amount dissolved rising with the temperature. In this respect solution differs from ordinary chemical action, which is not continuous. A substance, like common salt, may dissolve in water without any chemical action being manifest (although possibly it may take place), and a fall of temperature is noticed due to the latent heat of liquefaction of the solid. The liquid will spontaneously separate from the solid if exposed to the air, and the salt is recovered in its original state. If the solution of salt is made very cold however, a crystalline compound of salt with water can be obtained.

On the other hand, some salts, e. g. anhydrous calcium chloride, dissolve in water with the production of heat sensible to a thermometer; in such cases we have acting both the latent heat of liquefaction which would cause a fall of temperature, and a chemical union of greater energy which gives finally a positive rise, expressing the difference between the two. When the quantity of substance in solution is as great as the liquid can hold, the solution is *saturated*.

The following will serve as examples of solution:—

	100 parts of water dissolve.		Heat of solution.
Sodium Chloride	35.2 parts at	0°	− 1100 c.
Potassium Nitrate	13.3 ,,	0°	− 8300 c.
Calcium Chloride (dry)	63.3 ,,	10°	+ 9400 c.
Ferrous Sulphate, dry	27.1 ,,	10°	− 1300 c.
,, ,, crystals	61.0 ,,	10°	− 2300 c.
Lead Nitrate	38.7 ,,	10°	− 4100 c.
Potash	213.0 ,,	10°	+ 12460 c.

Many liquids mix with water in all proportions: alcohol, sulphuric acid and nitric acid are examples. In these cases we notice a rise of temperature and contraction of volume. There is doubtless chemical union taking place, but the mixing of the liquids is mechanical and differs from true solution, as the liquids will not separate as a salt does from water by change of temperature. The term saturated solution does not apply here. But again, ether and chloroform are soluble in water to a limited extent, dependent on temperature, and thus afford cases of simple solution of a liquid in water.

All gases are soluble, and the following table shows the greatly varying amounts which may be taken up by one volume of water:—

	Vols.
Hydrogen	.018 at $4°C$. and 760 mm.
Oxygen	.046 at $4°C$. ,,
Nitrogen	.018 at $4°C$. ,,
Chlorine	2.58 at $10°C$. ,,
Hydrogen Sulphide	4.23 at $2°C$. ,,
Ammonia	1180. at $0°C$. ,,
Hydrogen Chloride	503. at $0°C$. ,,

NATURAL WATERS.

The natural waters are sea water, river and lake water, spring water, well water, and rain water: and these differ as regards the kinds and quantities of substances dissolved in them. In sea water we have large quantities of saline matter (sodium, potassium, magnesium and calcium compounds): in river waters large or small quantities according to the character of the lands they drain. Spring waters of course vary widely in their dissolved matters, as they are derived from different geological strata. Rain water contains little but dissolved gases from the atmosphere, but in the neighbourhood of towns many impurities are washed out of the air by rain, and near the sea, in stormy weather especially, notable quantities of salt are found in rain water.

40   *Hardness of Waters.*

For household purposes waters are described as **hard** or **soft**, according as they require much or little soap to be used in washing with them. The **hardness of water** is most commonly due to compounds of lime and magnesia (see Calcium).

Distilled Water.

Since by evaporation water is separated from the fixed or non-volatile substances in solution we can get

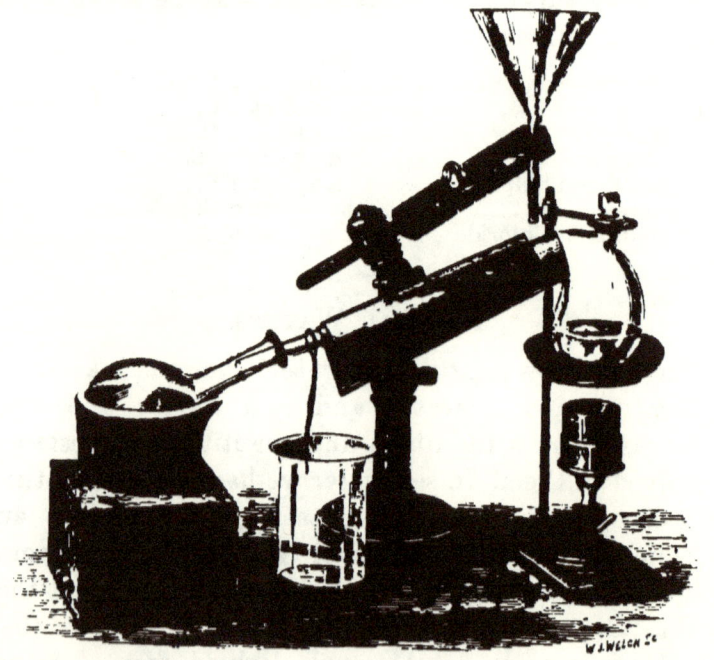

Fig. 20.

pure water for chemical purposes by distillation. An apparatus employed for the purpose is represented in the figure 20.

Or the water may be boiled in a tin can and the steam

led through a condenser or tube surrounded by a cold water jacket with a stream of water running through. As the steam condenses it drops from the end of the central tube. A condenser of this kind is shown in figure 46.

CHAPTER V.

Nitrogen and Air.
Nitrogen; N. Atomic weight, 14. Density, 14.
Molecular weight, 28 = N_2.

NITROGEN exists free in the atmosphere, of which it forms about four-fifths of the entire quantity. It is easily obtained from air by the use of any substance that can remove the oxygen by absorption or combination.

Fig. 21.

(1) A piece of phosphorus in a small dish standing in water is set on fire, and while burning covered with a bell jar (fig. 21). For a time the phosphorus continues to burn, but presently goes out before it is all consumed. The white fumes or smoke given off consist of oxide of phosphorus [P_2O_5]; these presently settle down and disappear, dissolving in the water to form phosphoric acid. As the gas left in the jar cools, the water rises within, and it will be found that about one-fifth of the volume of the air disappears:—the residual four-fifths are nitrogen. If tested by bringing a lighted paper into the jar, it will be seen that the gas immediately puts out the flame, since it cannot support combustion.

(2) A similar experiment, to show the quantity of nitrogen more exactly, is made by burning phosphorus in a long tube. A glass tube about thirty inches long and

half-an-inch bore is required; one end should be sealed, and the other closed with a sound well-greased cork (fig. 22). A rough graduation into five equal parts is made by slipping small rings of india-rubber tube upon the glass tube. A thin tube is less liable to break than a thick one, but to avoid risk of burns it is in any case best to hold the tube with a cloth. Let a piece of phosphorus be placed in the tube and the cork fitted tightly in its place: by bringing the tube over a lamp the phosphorus is melted and takes fire, when the tube should be inclined so as to cause the liquid-burning phosphorus to run to the bottom. All the oxygen in the tube is taken up, and if the tube is opened under water

Fig. 22.

it will be found that the residual *nitrogen* fills four-fifths of the tube and the water enters to the extent of one-fifth.

The oxygen of the air may be withdrawn by numerous methods: moist iron wire placed in a tube over water speedily absorbs the oxygen at ordinary temperatures: tin, mercury, copper, and other metals require to be heated.

The experiment with copper is made as follows. Some metallic copper is packed in a tube (a piece of iron gas-pipe serves well) and a stream of air passed over from a gas-holder (fig. 23, *a*): the tube containing the copper is placed in a gas-furnace (*d*) and heated to redness. The metal takes up the oxygen, and nitrogen passes on and can be collected over water in the jar (*e*) in a pneumatic trough.

$$Cu + O + x[N_2] = CuO + x[N_2].$$

Copper and oxygen form copper oxide; the nitrogen is unchanged.

A good method for obtaining nitrogen in quantity is to pass air through solution of ammonia, when it becomes charged with that gas, and then over red-hot copper as above. The copper oxide at first formed is continually reduced by the ammonia to the metallic state, and the

Fig. 23.

nitrogen thus set free passes on along with the atmospheric nitrogen:— .

$$2NH_3 + 3CuO = N_2 + 3Cu + 3H_2O.$$

Ammonia and copper oxide yield nitrogen, copper, and water.

The characters of the element nitrogen in the free state are chiefly negative; it is a gas without colour, taste, or smell. It will neither support life nor combustion, nor is it combustible itself. With lime-water it gives no turbidity. It cannot be absorbed in quantity by solvents, but it is slightly soluble in water. The spectrum of nitrogen obtained by the electric discharge in highly vacuous tubes is peculiar to itself, and affords one positive character of the element. In its compounds however, such as ammonia, nitric acid, etc., nitrogen shows extreme chemical activity.

AIR.

Atmospheric air is a mixture of gases, consisting for the most part of nitrogen and oxygen, of which the characters and properties have been described. In addition there is some carbonic acid gas (CO_2), about ·04 parts per hundred, or $\frac{1}{2500}$ on an average, and a very variable amount of moisture or vapour of water. The quantity of water is least when the temperature is very low, as in the cold air of Arctic Winter, and greatest where the temperature is high, as above the sea in summer within the Tropics.

The presence of carbonic acid in the air is shown by exposing lime-water in a dish, or better by bubbling air through clear lime-water in a bottle: by the union of the gas with lime a milky appearance is produced; the white compound formed being precipitated chalk:—

$$Ca(OH)_2 + CO_2 = CaCO_3 + H_2O.$$

The presence of water in the air is shown by cooling it. When warm moist air meets with cold air we observe the formation of clouds, mists, rain, snow, or hail; or the cooling of air near the ground may produce fog and dew. A vessel of cold water in a room, or glass windows, the day being cold outside, are often bedewed by condensed atmospheric vapour. We can also by artificial cooling as with (1) ice, or (2) ice and salt, or (3) the evaporation of ether, bring down the temperature of the air until dew is formed.

The quantity of carbonic acid in the air is found thus: From a measured volume of air the carbonic acid is absorbed and weighed. An aspirator is employed to measure and also to drive the air through (1) drying tubes and (2) absorbing tubes. For drying the air strong sulphuric acid is used (fragments of pumice being moistened with this liquid), and for absorption of the carbonic acid a tube filled with pieces of moist caustic soda or soda-lime is employed.

The soda tubes are weighed before and after the passing of the air, and the increase gives the weight of carbonic acid absorbed (fig. 24).

The quantity of water can be also determined by noting the increase in the weight of the drying tubes used in this experiment.

A rough estimation of the proportions of oxygen and

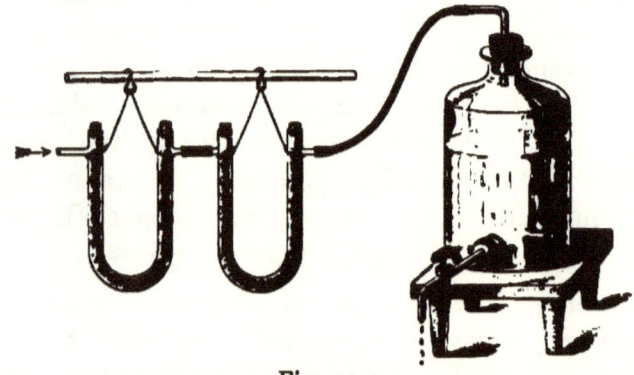

Fig. 24.

nitrogen in air is given by experiments already described, but an exact determination is a matter of great delicacy and needing many precautions to guard against error. The principles of the methods which have been used are as follows :—

DETERMINATION OF THE COMPOSITION OF AIR BY VOLUME.

It has been shown already that oxygen unites with exactly twice its volume of hydrogen to form water. Wherefore a mixture of measured quantities of air and hydrogen being exploded, and the amount of gas converted into water being measured, it is plain that one-third of the contraction is the volume of oxygen present in the air, and two-thirds are the volume of hydrogen used in combining with it.

(1) A eudiometer is used containing a convenient quantity of air over mercury; the volume of the air and the height of the mercurial column within the tube are read: readings of the barometer and the thermometer at the time of these measurements are also taken (fig. 17).

(2) Next some pure hydrogen (about an equal volume) is passed into the tube, and the volume of the gas and the length of the mercurial column of the eudiometer are again recorded.

(3) A spark is passed through the mixture, exploding it; and again the volume and mercury column are measured.

(4) The volumes thus recorded are reduced to standard pressure (760 mm.) and to the temperature of 0°C., and from these corrected volumes we may thus easily find the amounts of oxygen and nitrogen.

An experiment gave the following numbers:—

Volume of air employed, corrected to 0°C. and 760 mm. . . . . 455.
Volume after adding Hydrogen, corrected . . 689.
Volume after explosion, corrected . . . . . 403.

From these figures we get—

Contraction by explosion . . . . 286.
One-third of Contraction = Oxygen 95.3
Nitrogen by difference 359.7

  or N = 79.1 per cent.
      O = 20.9 per cent.

Determination of the Composition of Air by Weight.

The composition of air by weight was determined with great care by Dumas and Boussingault in 1841; their apparatus is represented in fig. 25, and consists of three portions:—

(1) A set of tubes for purifying the air, and containing potash to absorb carbonic acid and strong sulphuric acid to absorb moisture.

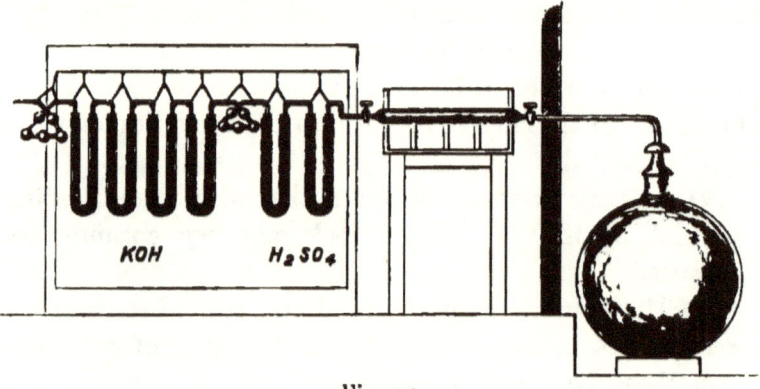

Fig. 25.

(2) A tube containing metallic copper, previously reduced in hydrogen gas, and furnished with stopcocks; this is placed in a heating furnace.

(3) A large glass globe with a stopcock which can be exhausted of air.

Before the experiment the tube containing copper and the globe are exhausted and weighed: the apparatus is then joined together and the tube containing the copper heated. On slightly opening the taps, air enters the open end of the apparatus, passing over potash and sulphuric acid: thus purified it enters the hot tube, where the oxygen is taken up by the copper and pure nitrogen passes on, into the vacuous globe. When the globe is full of nitrogen, the taps are closed and the apparatus is cooled and weighed.

The globe is heavier by the quantity of nitrogen it contains, and the tube is heavier by the oxygen combined with the copper. Also as the tube contains a little nitrogen after the experiment, it must be exhausted and weighed again; the little nitrogen being added to the first weight. The result of the experiment is—

    Oxygen . . 23.0 per cent. by weight.
    Nitrogen . . 77.0 per cent.

It is obvious that by comparing the weighings of the globe (1) filled with nitrogen, (2) empty, (3) filled with dry air, and supposing the pressure and temperature constant, the **density** of nitrogen may be found. And by refilling with pure oxygen and weighing, the **density** of oxygen can be found. The experiment gave—

    Density of Air = 1.
    Density of Nitrogen = .972.
    Density of Oxygen = 1.057.

The gaseous constituents of the air are merely in a state of mixture and not in chemical combination. This is proved in the following way:—

(1) The ratio of the weights of oxygen and nitrogen is not a simple multiple of their atomic weights.

(2) When oxygen is mixed with nitrogen, no sign of chemical action is observed: there is no change in the volume or temperature of the mixture.

(3) When air is dissolved in water, the oxygen being more soluble than nitrogen, a larger relative quantity of oxygen is dissolved; while the proportions of a compound gas are not altered by solution.

(4) An artificial mixture of oxygen and nitrogen, in the same proportions as the constituents of air, behaves exactly like air in all respects.

CHAPTER VI.

Carbon. C. Atomic weight, 12. Molecular weight, unknown.

THE element **Carbon** is one of great interest and importance as it forms a large proportion of all living animals and plants. Our food is all carbonaceous: flesh and albumen are composed chiefly of carbon, hydrogen, oxygen, nitrogen; fat and all starch foods contain the elements carbon, hydrogen, and oxygen. We burn carbon and substances containing it, to produce heat and light; and the carbon thus used as fuel is obtained, directly or indirectly, from animals and plants; for example tallow, wax, wood, coal, fats, oils and gas, are all of animal or vegetable origin.

The commonest forms of carbon are **soot, charcoal** and **coke**; besides these are the natural forms **graphite** (or black lead) and the **diamond**.

The **Diamond** is nearly pure crystallized carbon; it is the hardest known substance, and when cut and polished is brilliant and beautifully transparent with a powerful refractive effect upon light. The density of this form of carbon is 3·5.

Graphite or black-lead is a dark-grey mineral substance in scaly plates composed chiefly of carbon. It is used for making drawing pencils, and on account of its semi-metallic lustre for polishing stoves; a coating of graphite on gunpowder preserves it from damp. The density is about 2·3.

Charcoal is a form of carbon obtained by heating organic matter to full redness out of contact of air.

Forms of Carbon.

Animal charcoal is obtained by heating hide, leather, lood, bones, etc., in close retorts. It is chiefly used to remove the brown colour from sugar-solutions.

Wood charcoal is made by burning wood in heaps from which the air is kept out by a covering of turf only a few small holes being left. The wood contains carbon, hydrogen and oxygen, and if ignited with free access of air burns away altogether leaving only a little ash; but when burnt with a very limited supply of air the hydrogen and oxygen are driven off, and carbon is left as charcoal, retaining the form of the original wood.

Coke is obtained from coal, heated in close vessels as in the ordinary process of gas-making. A better kind is prepared from coal heated in ovens made for that express purpose. The volatile portions of the coal are driven off by heat, and a porous mass of carbon remains.

Soot is deposited during the incomplete combustion of various substances such as gas, tallow, oil, etc. When an insufficient quantity of air reaches the flame of a lamp, the combustion is not complete, but smoke is given off containing much carbon which deposits in the form called soot. A special kind of soot or **lamp black**, made from resin and similar bodies, is used in the manufacture of printer's ink.

We see thus that carbon is known in three distinct natural forms: (1) in diamond as a regular crystal; (2) in graphite as crystalline plates; (3) in the uncrystallized (amorphous) states of charcoal, coke, soot. It is proved that these are different forms (*allotropic* forms as they are termed) of the same element by burning them. For each form of carbon produces when burnt in oxygen the same gas—carbonic acid gas CO_2, and from equal weights of carbon in either form we find that equal weights of carbonic acid are obtained. That is to say, 12 parts by weight of carbon from any source combine with 32 parts by weight of oxygen and always yield 44 parts by weight of carbonic acid gas.

$$C + O_2 = CO_2.$$
$$12 + 32 = 44.$$

Carbon is in all its forms an infusible substance: it burns in air or oxygen to form the dioxide (CO_2) except in special cases when the monoxide CO is obtained.

A remarkable property of wood charcoal is the power it has of absorbing gases of all kinds, and, more especially condensible gases such as ammonia, cyanogen, hydrogen sulphide, etc. This fact can be shown by collecting ammonia gas in a tube over mercury, and passing into it a piece of freshly heated boxwood charcoal (fig. 26): the whole of the gas, if pure, will be absorbed in a few minutes.

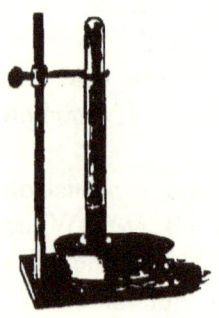

Fig. 26.

Or a piece of charcoal is first allowed to absorb hydrogen sulphide, and then brought into a bottle of oxygen; the oxygen is condensed, and combines with the hydrogen and sulphur with avidity, so that the charcoal will sometimes take fire. This property of charcoal can be made use of for absorbing offensive gases from drains, which are not only absorbed, but in fact burnt up by being caused to combine with oxygen of the air.

Animal charcoal is especially remarkable as an absorbent for organic colouring matters. Thus if a dilute solution of indigo be passed through a tube packed with this substance the liquid will become perfectly colourless. Accordingly this form of charcoal is used as a filtering material in a variety of laboratory operations, and in the arts, especially for decolourising syrups in the manufacture of white sugar from brown.

For the purification of drinking water, filters of animal charcoal are very commonly employed.

The diamond is a non-conductor of electricity while the graphitic forms of carbon conduct more or less. But compared with metals the resistance of carbon conductors is large, and they therefore become strongly heated by a powerful electric current. On this account rods of a dense form of coke are used for electric arc lights, and the filaments of glow or incandescent lamps are made of carbonised fibres.

Oxides of carbon. When charcoal or any form of carbon burns in the air, a compound with oxygen is formed called **carbon dioxide** or carbonic acid gas. Let a piece of charcoal be ignited and plunged into oxygen in a deflagrating jar (fig. 10) it glows and burns brilliantly. The formation and presence of carbonic acid is shown by pouring a little clear lime-water into the bottle which becomes white and milky from the precipitation of chalk or carbonate of lime:—

$$Ca(OH)_2 + CO_2 = CaCO_3 + H_2O.$$

To prepare the gas in a pure state we decompose some carbonate with an acid; the carbonate of lime is commonly used in the form of **chalk** or **marble**.

Some fragments of marble (or chalk, or any convenient carbonate) are placed in a flask (fig. 27) with a little water and hydrochloric acid poured in through a thistle funnel: the gas comes off readily without any heating.

The chemical changes are:—

$$CaCO_3 + 2HCl = CaCl_2 + H_2O + CO_2.$$

Carbon Dioxide.

The following experiments will show the properties of the gas:—

1. It can be collected in jars by downward displacement showing that it is heavier than air.

2. A burning taper is extinguished if plunged into the gas.

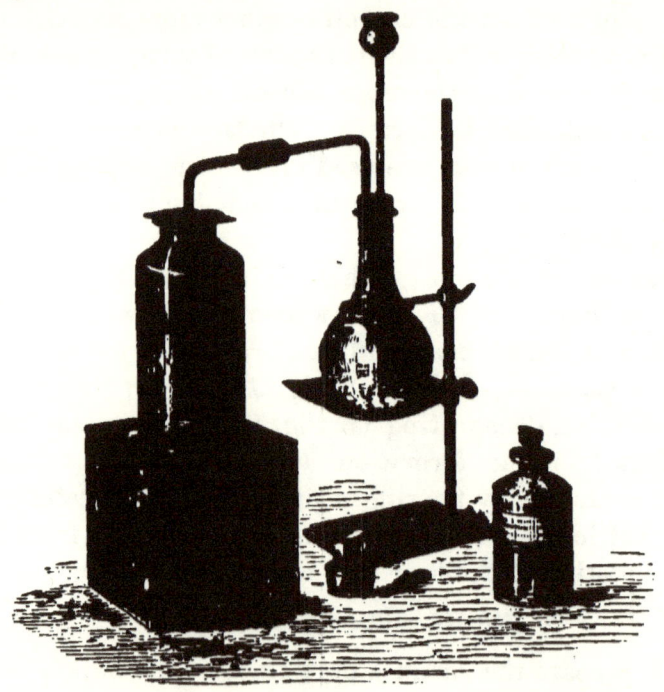

Fig. 27.

3. Some clear lime-water poured into the gas becomes white and milky from formation of carbonate of lime.

4. Pour a jar of the gas into an empty jar and test with a taper or lime-water, to show that the gas has passed into the jar.

5. Although the gas does not support combustion of a taper it will give up its oxygen to certain metals.

Thus a piece of magnesium ribbon will burn in it, and if the white oxide (magnesia) produced is dissolved in a little hydrochloric acid, fragments of solid carbon will be left undissolved, having been separated from the gas (fig. 11):—

$$CO_2 + 2Mg = 2MgO + C.$$

6. Allow the gas to pass into water: some will be dissolved, and the solution will become faintly acid to litmus paper, and will also give a white precipitate with lime-water.

If a stream of the gas be passed through the liquid, this precipitate will dissolve, leaving the solution clear; this change is due to the formation of a soluble bicarbonate of lime (see Calcium).

7. Allow the gas to pass into a strong solution of soda (or potash), it will at first be entirely absorbed, and carbonate of soda will form in the liquid:—

$$2NaOH + CO_2 = Na_2CO_3 + H_2O.$$

If the action of the gas is continued, a bicarbonate or acid carbonate is formed:—

$$2NaOH + 2CO_2 = 2NaHCO_3.$$

The combustion of carbon (as charcoal) in oxygen has been found to give out 8080 units of heat for each gramme of carbon; or 12 grammes of carbon with 32 grammes of oxygen will produce 96960 units of heat.

Properties of Carbon Dioxide.

Carbon dioxide is a gas, colourless, transparent and invisible under ordinary conditions, but it may be condensed to a liquid if pumped into a strong iron vessel. When the liquefied gas is allowed to escape into the air, it expands very quickly, and so large is the quantity of heat absorbed by being converted into work in the expansion that a great fall of temperature takes place, and part of the liquid is frozen. Solid carbon dioxide thus obtained is a light white substance resembling snow, and cold enough to freeze mercury.

Carbonic acid gas is soluble in water, which at 15°C dissolves its own volume; the solution has a slightly acid taste; under pressure a large amount of the gas is taken up. The common aërated waters, so called, such as soda water and seltzer water, are charged with this gas by pumping it into strong vessels containing the saline waters which afterwards are drawn off into bottles strong enough to withstand the pressure. Carbonic acid is formed in large quantity during fermentation. The cavities in fermented bread are produced by the liberation of carbon dioxide during the fermentation produced by the growth of the yeast plant; and similarly this gas is liberated in the fermentation of beer, wine, and all alcoholic liquids. The splitting up of sugar and formation of alcohol and carbonic acid takes place thus:—

$$C_6H_{12}O_6 = 2C_2H_6O + 2CO_2.$$

Glucose produces alcohol and carbon dioxide.

It has been seen that carbon dioxide does not support combustion, neither does it support life, animals being suffocated in the gas. Accidents frequently happen owing to an accumulation of this gas in old wells; and workmen who descend without precaution becoming insensible in the vitiated air, die before rescue is possible. The air of a well can easily be tested by lowering a lighted candle before anyone descends: if the flame is extinguished the air is of course bad.

The air expelled from the lungs of animals in breathing contains a considerable proportion of carbonic acid gas.

Carbon Dioxide.

To show this fact blow gently from the mouth through a glass tube into lime-water, and observe how soon a white precipitate of carbonate of lime is formed. The gas thus given off ought not to be allowed to accumulate in rooms, but a continual change of air should be provided by proper **ventilation**. Plants in a similar way by respiratory processes give off carbonic acid from their leaves.

But on the other hand the green leaves of plants (containing chlorophyll) serve to purify the air by removing carbonic acid from it. In the nutrition of the plant this gas is taken up by leaves and the carbon retained, while the oxygen combined with it is returned to the atmosphere. So in the economy of nature the life and growth of plants, broadly speaking, has an opposite effect to the life of animals, by restoring to the air the oxygen which previously had been consumed by the latter.

COMPOSITION OF CARBON DIOXIDE.

The quantities of carbon and oxygen which combine to form this oxide are found by burning a weighed portion of carbon in a stream of oxygen and weighing the product: the difference in the two weights gives the oxygen consumed.

The experiment is performed as follows:—

Pure oxygen stored in the gas holder (fig. 28) is passed through drying tubes containing (1) potash and (2) sulphuric acid (the last of which is weighed to test and prove by its constant weight the dryness of the gas); it then passes into a glass or porcelain tube kept at a red heat by a furnace. In this tube is the carbon, in the form of diamond or graphite, which has been carefully weighed. The carbon burns and the oxide formed passes on into the weighed absorption tubes, the first of which contains sulphuric acid to retain any water which may possibly be formed if the substance burnt should contain any hydrogen. The other tubes are filled with potash to absorb and retain the carbon dioxide; these are carefully weighed before and after the experiment. By this means the weight of carbon burnt and the quantity of the dioxide produced are accurately known, and the composition of the oxide can be calculated as follows:—

Weight of carbon burnt . . . . = 1. grammes.
Increase in potash tubes [due to carbon dioxide] . . . . . = 3.666 „
Difference (oxygen) . . . = 2.666 „

From which, if we suppose 12 parts of carbon to be burnt,

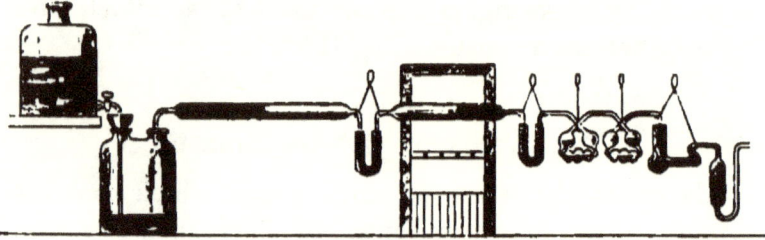

Fig. 28.

we see that 32 of oxygen are used, making 44 of carbon dioxide. The density of the gas is one half the weight of the molecule CO_2, viz. 22 ($H=1$).

$$\frac{12 + 32}{2} = \frac{44}{2} = 22.$$

11.2 litres of hydrogen weigh 1 gramme.
22.4 „ oxygen „ 32 grammes.
22.4 „ carbon dioxide „ 44 „

It can be shown by an experiment that the volume of oxygen is not altered by burning carbon in it. A piece of charcoal is supported by a wire in a bulb containing oxygen gas (fig. 29): by concentrating sunlight on it with a large lens the charcoal is ignited (or an arrangement can be made for firing the charcoal by electricity), and after the whole has cooled the quantity (volume) of gas is found to be unaltered.

Carbon monoxide or **carbonic oxide**, CO. This oxide is formed by the imperfect combustion of carbon in air or oxygen, but cannot be obtained pure in that manner.

If a current of carbon dioxide be slowly passed through an iron tube packed with charcoal and heated to redness in a furnace, it is reduced to the monoxide and doubles its volume:—

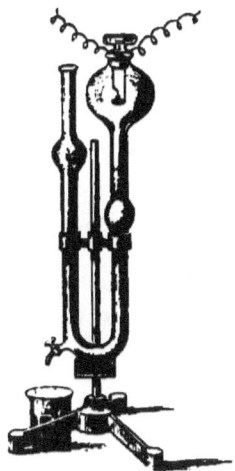

Fig. 29.

$$CO_2 + C = 2CO.$$
2 vols.   4 vols.

If steam in a similar way is sent over the charcoal we have the reaction—

$$H_2O + C = CO + H_2$$

a mixture of carbonic oxide with hydrogen being obtained, but some dioxide is always produced according to the equation:—

$$H_2O + CO = CO_2 + H_2.$$

This oxide is combustible and burns in air with a blue flame, taking an additional atom of oxygen to form dioxide:—

$$2CO + O_2 = 2CO_2.$$

When a coke or charcoal fire is burning we get both oxides formed: in the lower part of the fire where the supply of oxygen is largest the combustion is complete, and carbon dioxide is formed, but in rising through the red-hot fuel this is in part converted into monoxide. And at the top of a clear fire the monoxide gets a further supply of oxygen and may be seen burning with a pale blue flame.

A good method for preparing carbonic oxide in the laboratory is to heat crystals of oxalic acid with strong sulphuric acid.

$$H_2C_2O_4 + H_2SO_4 = (H_2SO_4 + H_2O) + CO + CO_2.$$

The oxalic acid breaks up into water (which combines with the sulphuric acid) and a mixture of the two oxides

Fig. 30.

of carbon. By passing these gases through a solution of soda the dioxide is absorbed, forming carbonate of soda, and the monoxide passes on; it is collected over water.

Carbon monoxide is an invisible colourless gas. It is very poisonous, and dangerous therefore to breathe. Unlike the dioxide it does not give any precipitate with lime-water. Cuprous chloride in solution (in hydrogen chloride, etc.) absorbs the gas entirely. Its density is $\frac{12+16}{2} = 14$.

The apparatus for preparing the gas is shown in the figure 30. A few jars of gas having been collected, the following experiments may be performed.

(1) Shake up a little of the gas with lime-water: no precipitate will form if the gas is pure. If any carbonic acid is present it can first be absorbed by a little soda solution brought into the jar.

(2) Inflame a portion of the gas and observe that it burns with a pale blue flame; then test the burnt gas with lime-water, which will show the presence of carbon dioxide.

(3) A mixture of carbonic oxide with half its volume of oxygen will explode on the application of a flame.

Combustion. The combustible substances in common use for producing heat and light contain a large proportion of carbon, as is shown in the following examples:—

	Carbon per cent.
Charcoal	93.
Coke	88.
Anthracite Stone Coal	90.
Coal	80.
Tallow	76.
Wax	80.
Paraffin	84.

The products of the complete combustion of bodies containing carbon and hydrogen, as in the flame of a candle, or gas burner, are carbonic acid gas and water; the presence of the former can be shown by the precipitate produced with lime-water; and of the latter by the dew deposited on any cold object brought into the flame.

Structure of Flame. The flame of a candle (fig. 31) is seen to consist of three distinct portions :—

Structure of Flame.

(*a*) The gaseous cone containing unburnt combustible vapour.

(*b*) The bright zone of active combustion.

(*c*) The dim mantle of hot gases.

The interior of the flame is composed of hot gases: the combustible vapours from the heated tallow. No combustion can take place here, as no oxygen can reach the fuel: but if a small glass tube be held in the hollow space the gases will pass through it, and may be lit at the distant end.

In the second zone combustion in oxygen is rapidly

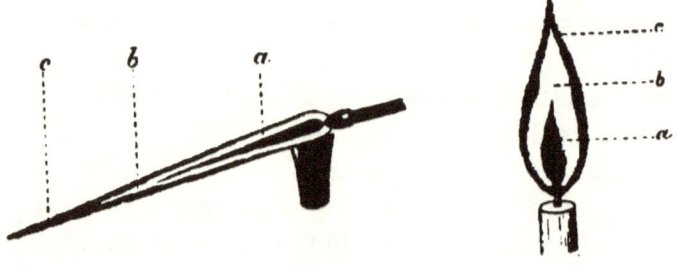

Fig. 31.

taking place: the combustion beginning in the centre and becoming complete on the outside. The luminosity of the candle is entirely due to this portion of the flame, and depends on the presence of incandescent particles of carbon. If a cold object is brought into this zone the carbon is deposited in the solid form as soot.

The outer mantle, which is scarcely luminous, contains heated gases, carbonic acid, and water, from the flame, and nitrogen and oxygen from the air.

The flame of a blowpipe is precisely similar in character: but being better supplied with oxygen the combustion is more rapid and a higher temperature is reached. The zone *a* is a moderately cool reducing zone:

b is a hot reducing zone, and the point of the cone is the hottest spot in the flame. The mantle *c* is blue, and its tip is the best oxydising flame (see fig. 31).

The flames of gas burners of the common type will be seen to be also precisely similar in structure.

CHAPTER VII.

Sulphur. S. Atomic Weight, 32.

THE element sulphur is found in small quantity in the free state near volcanoes in Sicily and other places, and has long been known to man; but for the most part sulphur is in nature associated with metals in the form of sulphides or sulphates. The commonest sulphides are iron pyrites FeS_2, copper pyrites $[FeCu]S_2$, and galena or lead sulphide PbS: and among common sulphates we have sulphate of lime $CaSO_4$ in the forms of gypsum, or alabaster, or plaster of Paris; the sulphates of the earthy metals barium and strontium being less frequently met with.

Where native sulphur is procurable it is separated from earthy matter by heat, either by simple fusion or distillation. The melted sulphur, roughly cast into stout rods or sticks, is commonly known as roll sulphur or **brimstone**. A purer form is obtained by sublimation; the vapour from boiling sulphur is brought into a large walled chamber, in which it condenses to the powdery solid known as **flowers of sulphur**.

Iron pyrites heated in a closed tube gives off about one-third of its sulphur, and in practice this substance is sometimes roasted in a limited supply of air to obtain sulphur from it:—
$$3FeS_2 = Fe_3S_4 + S_2.$$

A certain quantity of sulphur also is obtained from spent oxide of iron used in purifying coal gas, and methods have been devised for recovering sulphur from the 'alkali waste' produced in the soda manufacture.

Common sulphur, as it appears in the form of brimstone, is an opaque yellow brittle substance, almost tasteless, but with a slight unpleasant odour. It conducts heat badly, and splits and crackles if suddenly brought into a flame: it is also a bad electrical conductor, and becomes easily electrified by friction.

Allotropic forms. Sulphur is remarkable among elements for the number of physical forms in which it appears and the ease with which they change from one to another.

Octahedral sulphur is the form in which natural sulphur is found and the form into which apparently all others have a tendency to change. Native sulphur is a permanently transparent yellow solid, crystallised in octahedra, and quite stable in character. Roll sulphur and flowers of sulphur consist chiefly but not entirely of this variety. Artificial crystals may be obtained by digesting roll sulphur in warm carbon disulphide, filtering and allowing the liquid to cool and crystallise (fig. 32).

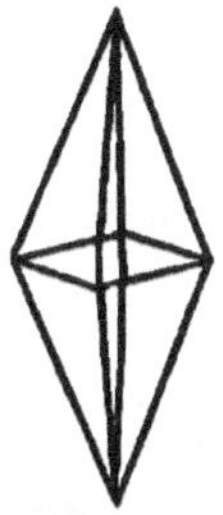

This form is readily soluble in carbon disulphide; its melting point is 114.5 C. and its specific gravity is 2.05.

Prismatic sulphur is obtained by melting ordinary sulphur and allowing it to cool until it becomes partially solidified. About half a pound or a pound of sulphur is cautiously melted in a hemispherical dish until it just becomes liquid; the temperature should not rise much above 120° or the sulphur will pass into a viscid condition: at the right heat the sulphur forms a thin mobile orange liquid. The lamp being removed the sulphur is allowed to cool until a

Fig. 32.

crust of crystals forms on the top: as soon as the top is solid a couple of holes should be made in it with a rod, and the portion of the still liquid sulphur inside quickly poured out. If the crust be then broken away the interior will be found full of the needle-shaped crystals of prismatic sulphur (fig. 33).

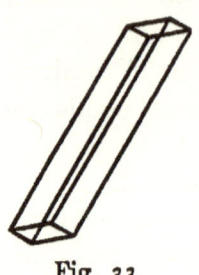

Fig. 33.

The prisms are at first transparent, but cannot be kept, since after a few hours they become yellow and opaque, breaking up into the common form of sulphur, and the needles become changed into collections of minute octahedra.

The melting-point of this variety is 120° C. and its specific gravity 1·98.

Plastic sulphur. If a quantity of sulphur in a flask or test-tube is more strongly heated, it will be seen to darken in colour and become thick and treacly. This viscid condition is most marked about the temperature 240°, and in this condition the sulphur can hardly be poured out of the tube. If the mass be cooled suddenly in water, the plastic state is retained for some days. It has been found that while this change is being produced the temperature of the sulphur is almost constant, and a thermometer placed in it rises very slowly; the heat of the flame being absorbed in doing internal work producing molecular change.

Distillation of sulphur. A quantity of sulphur is heated in a glass retort until the temperature rises to the boiling-point 440°, when it is converted into a darkbrownish vapour. By adjusting the heat properly, the vapour can be condensed in the neck of the retort, and a clear stream of liquid sulphur will trickle from the end and if received in a beaker of water (fig. 34) form transparent elastic threads. The plastic state lasts but a few hours.

In addition to the forms above described there are less distinct varieties of an amorphous or non-crystalline form of which some are soluble and others insoluble in bisulphide of carbon, etc.

The vapour density of this element is also remarkable and shows that two forms of sulphur vapour exist.

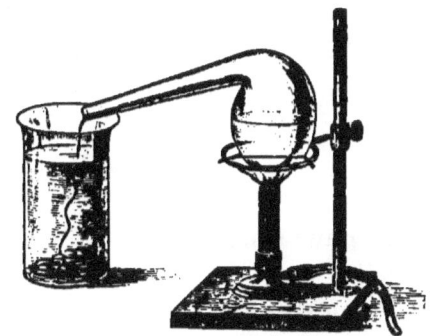

Fig. 34.

The density of the vapour at a temperature of 500° is 96; but at much higher temperatures, at 1000° for example, the density is 32 in relation to hydrogen as unity. Thus the quantity of sulphur in vapour which fills the space of two atoms of hydrogen [H_2] is 192 or [S_6] in the first case, but in the second H_2 occupies as much space as 64 of sulphur or [S_2]. So that if a molecule of hydrogen is represented by the symbol H_2, we have a molecule of sulphur represented by S_6 or S_2 according to the temperature. How many atoms of the element unite to form the molecules of the solid varieties, we have no means at present of discovering. It has been shown that the conversion of octahedral sulphur into the other varieties is attended by an absorption of heat (which disappears and becomes latent while doing internal molecular work) and similarly in the reverse changes, heat is given out which is manifested by a rise in temperature when either of the varieties changes back into the octahedral form.

All the forms of sulphur are combustible, burning with a blue flame, and yield identical products when burnt in air or oxygen.

The following are some of the most important compounds of sulphur:—

SO_2 sulphur dioxide.
SO_3 sulphur trioxide.
H_2SO_4 sulphuric acid.
H_2S hydrogen sulphide.
CS_2 carbon disulphide.

Sulphur dioxide, SO_2. Molecular weight, 64; vapour density, 32.

When sulphur is set on fire it burns with a blue flame giving off a gas with a well-known pungent odour; and whether the combustion is in air or oxygen the only product appears to be the dioxide (SO_2).

The combustion in oxygen takes place without any alteration in volume and this is shown by the apparatus used to prove a similar relation in the case of carbon (fig. 29):—

$$S + O_2 = SO_2.$$
$$\text{2 vols.} \quad \text{2 vols.}$$

This gas, also known as sulphurous anhydride, is usually prepared from sulphuric acid by acting upon it with copper. Some pieces of sheet copper are placed in a flask and covered with concentrated sulphuric acid: it is necessary to use the strong acid for this purpose. Upon applying a gentle heat the gas is readily given off, and being very heavy can be collected in jars or bottles (fig. 27), or as it is soluble in water it can be collected over mercury:—

$$Cu + 2H_2SO_4 = SO_2 + CuSO_4 + H_2O.$$

Copper and sulphuric acid yield sulphur dioxide, copper sulphate and water.

Sulphurous Acid.

Sulphuric acid itself can be decomposed at a red heat by being slowly dropped on hot bricks, etc., giving sulphur dioxide, oxygen, and water:—

$$H_2SO_4 = SO_2 + O + H_2O.$$

Sulphur dioxide is a transparent colourless gas with a pungent acid taste and smell, characteristic of burning sulphur. It easily liquefies and upon simply leading the gas into a glass tube cooled in a freezing mixture of ice and salt, it condenses to a colourless liquid which may be preserved by sealing the tubes.

When the gas is led into water a large quantity (about 50 vols. at 10°) is dissolved and a strongly acid liquid obtained, which is a solution of **sulphurous acid**, $SO_2 + H_2O = H_2SO_3$, by cooling the solution crystals may be obtained $[SO_2, 15 H_2O]$.

Although this acid cannot be isolated in a definite form, if neutralised with alkaline hydrates or carbonates crystallised salts are obtained. Thus

$$2NaOH + H_2SO_3 = Na_2SO_3 + 2H_2O.$$
$$NaOH + H_2SO_3 = NaHSO_3 + H_2O.$$

One molecule of acid with two of sodium hydrate yields the neutral sodium sulphite, while one molecule of acid and only one of sodium hydrate form the acid sodium sulphite.

The sulphites are decomposed with effervescence upon the addition of strong acids in a manner similar to carbonates.

$$Na_2SO_3 + 2HCl = 2NaCl + SO_2 + H_2O.$$

Sodium sulphite and hydrogen chloride yield sodium chloride, sulphur dioxide and water.

Sulphur dioxide is sometimes employed for bleaching silk, wool, straw, etc.: and the fumes of burning sulphur having a powerful antiseptic action are frequently used for disinfecting sick rooms and ships.

Thiosulphates. When sodium sulphite is digested with sulphur it is converted into sodium thiosulphate:—

$$Na_2SO_3 + S = Na_2S_2O_3.$$

This salt, better known as **hyposulphite**, is largely used in photography as a solvent for silver chloride, bromide and similar compounds.

Sulphuric acid, H_2SO_4. Molecular weight, 98.

This acid, commonly termed 'Oil of Vitriol,' is on account of its great usefulness for manufacturing purposes the most important of sulphur compounds.

It is prepared from the lower oxide of sulphur by the addition of oxygen and water—

$$SO_2 + O + H_2O = H_2SO_4.$$

On the small scale in the laboratory, this oxidation is brought about in various ways.

(1) A solution of sulphurous acid slowly absorbs oxygen from the air.

$$H_2SO_3 + O = H_2SO_4.$$

(2) A solution of sulphurous acid is treated with chlorine water—

$$H_2SO_3 + Cl_2 + H_2O = H_2SO_4 + 2HCl.$$

or any substances which readily part with oxygen may be used, such for example as nitric acid, chromates, permanganates, etc.

But for the manufacture of the acid on the large scale, the oxidation is effected by atmospheric oxygen through the agency of oxide of nitrogen; the process adopted being briefly as follows:—

Iron pyrites (or sulphur) is burnt in a brick chamber (fig. 35, *A*), and the products of combustion, viz. sulphur dioxide, mixed with nitrogen and some oxygen, conveyed by a large flue passing through a Glover's tower (B) into the first of a series of spacious chambers. These chambers ($C_1 C_2 C_3$) are constructed upon wooden frames, lined throughout with sheet lead. The diagram shows the general arrangement of the apparatus.

In this 'lead chamber' the conversion of the sulphurous gas into sulphuric acid, by the joint action of oxygen and steam, is effected. On the way to the

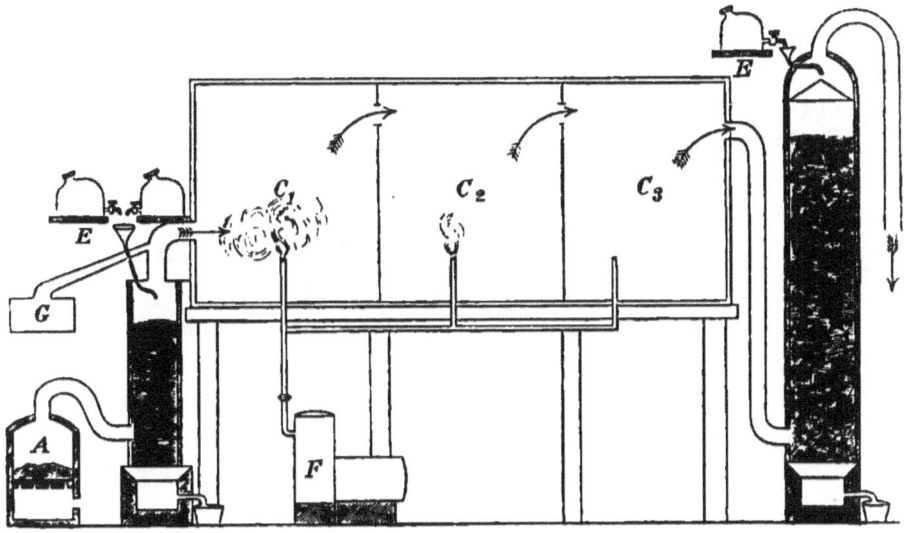

Fig. 35.

chamber the heated gases are allowed to mix with nitric acid vapours, and jets of steam from F are admitted.

The changes which follow are expressed by the equation

$$SO_2 + NO_2 + H_2O = H_2SO_4 + NO.$$

Sulphur dioxide, nitrogen peroxide, and water, yield sulphuric acid and nitric oxide. The latter gas has the property of combining with free oxygen to form the peroxide, so that upon a suitable quantity of air being admitted to the chambers the peroxide of nitrogen is formed afresh as fast as it is destroyed.

$$2NO + O_2 = 2NO_2.$$

This again comes into contact with more sulphurous gas oxidizing a fresh portion, and so on continuously: a small quantity of nitrogen peroxide thus suffices by alternate reduction (or de-oxidation) and oxidation, to convert a large quantity of sulphur dioxide into sulphuric acid.

The necessary nitric acid vapours are produced in the first instance, by acting on nitrate of soda with sulphuric acid; the pots (G) containing the mixture being usually placed so as to get heated by waste heat from the pyrites burners.

The nitrogen contained in the atmospheric air consumed in the oxidation of the sulphur in both stages of the manufacture, must be removed from the chambers. This is effected by means of the draught of a large chimney; and the gases of the lead chambers are drawn gradually out from the end distant from the furnace. But in order to catch any nitric oxide which would otherwise pass away and be lost in the air, a tower, called a **Gay-Lussac's** tower, marked D in the diagram, filled with coke is interposed, and the gases from the chambers are passed through it. The fragments of coke in the Gay-Lussac or denitrating tower are saturated from E with strong sulphuric acid which absorbs nitric oxide vapours, and the gases when thoroughly scrubbed pass out into the atmosphere.

This nitrated acid is again utilised by being sent down the tower called a **Glover's tower**, which is erected between the pyrites burner and the first lead chamber. Here it, after some dilution with weaker acid contained in E, meets the furnace gases and gives up its dissolved nitric oxides, which mixing with the sulphurous gases pass onward into the lead chamber. The loss by waste of oxides of nitrogen when this arrangement is carefully worked, is reduced to a very small quantity.

The sulphuric acid formed in the chambers settles down upon the floors, and is drawn off from time to time. It is still weak, containing about 64% of acid, and must be concentrated by evaporation of the water by heat.

Pure acid for laboratory purposes is distilled in platinum stills.

The commercial acid is known by the name 'oil of vitriol,' it is frequently slightly brown from traces of organic matter, and it may contain the following impurities, (1) lead derived from the chambers, (2) arsenic from the pyrites used, (3) traces of nitric acid. The strongest acid has a density of 1·85, and is a heavy colourless oily liquid. It has a great affinity for water, and is used greatly in the laboratory for drying gases. Much heat is given out when the acid is mixed with water, on which account the dilution of the strong acid must always be done with caution, the acid being slowly poured into the water.

Sulphuric acid is very corrosive, and rapidly attacks and destroys organic substances. To illustrate the attraction of the acid for water, a syrupy solution of cane sugar is mixed with the strong acid; it decomposes with energy, and leaves a mass of porous carbon similar to that left by heating sugar.

$$C_{12}H_{22}O_{11} = 12C + 11H_2O.$$

Diluted sulphuric acid has a sour taste, turns litmus paper red, and decomposes carbonates with effervescence.

If the acid be added to a solution of alkaline carbonate, or hydrate of soda, potash, or ammonia, until the liquid is neutral to litmus paper, a salt is formed in solution which by evaporation on a water bath can be obtained in crystals: such a salt is a neutral sulphate.

$$K_2CO_3 + H_2SO_4 = K_2SO_4 + CO_2 + H_2O.$$

Potassium carbonate yields potassium sulphate. If exactly twice the quantity of acid is used, and the solution evaporated, the salt obtained is the acid sulphate or potassium bisulphate.

$$K_2CO_3 + 2H_2SO_4 = 2KHSO_4 + H_2O + CO_2.$$

In the first instance the whole of the hydrogen of the acid is replaced by two atoms of the metal potassium, while in the second case only one half the hydrogen is so replaced.

When the hydrogen of an acid is divisible by two, and can be displaced by bases in two stages, the acid is termed **di-basic**.

Sulphuric acid, sulphurous acid, and carbonic acid, are examples of **dibasic** acids, inasmuch as they form both 'neutral' salts and 'acid' salts.

The following salts will serve as examples:—

Na_2SO_3 sodium sulphite : neutral.
$(NH_4)_2SO_4$ ammonium sulphate : neutral.
K_2CO_3 potassium carbonate : neutral.
$NaHSO_3$ sodium bisulphite : acid.
$KHSO_4$ potassium bisulphate : acid.
$NaHCO_3$ sodium bicarbonate : acid.

Sulphuric anhydride or Sulphur trioxide, SO_3.

This oxide is not formed by burning sulphur in oxygen, but can be made by passing a mixture of sulphur dioxide with oxygen over heated platinum sponge or platinised pumice (fig. 36).

Sulphur Trioxide.

Under these conditions the gases will combine with the formation of white clouds of sulphuric anhydride. For manufacturing purposes the mixed gases are obtained from strong sulphuric acid by dropping it gradually into a red hot (platinum) vessel, when it splits up as follows

$$H_2SO_4 = SO_2 + O + H_2O,$$

and the water being separated by cooling and drying the gases, a mixture of the two constituents is obtained

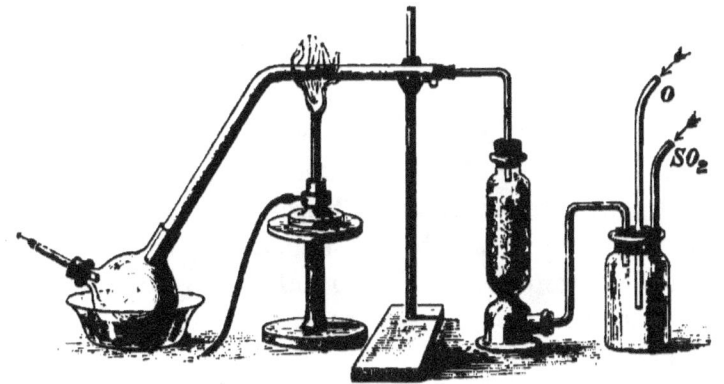

Fig. 36.

in the proportion desired for the subsequent conversion into sulphuric anhydride.

Sulphuric anhydride combines with water with much energy to form sulphuric acid—

$$H_2O + SO_3 = H_2SO_4$$

and also with sulphuric acid to form a crystallizable fuming acid compound—

$$H_2SO_4 + SO_3 = H_2S_2O_7.$$

The fuming sulphuric acid, otherwise known as Nordhausen sulphuric acid, easily parts with the combined anhydride which distils off by a gentle heat, and thus furnishes a convenient source for the preparation of small specimens of the anhydride.

Sulphuric trioxide is a white fusible solid crystallizing in needles or prisms.

Hydrogen Sulphide. H_2S. Molecular Weight, 34. Vapour density, 17.

This gaseous compound of hydrogen and sulphur, is commonly prepared by acting on ferrous sulphide (iron sulphide) with sulphuric or hydrochloric acid.

$$FeS + 2HCl = FeCl_2 + H_2S.$$

Other metallic sulphides similarly treated will yield the gas.

Hydrogen and sulphur will combine directly on being heated, and nascent hydrogen also reacts with sulphites or sulphurous acid to form this sulphide. Thus, if a little solution of a sulphite is added to a tube in which hydrogen is generated in the usual way by hydrochloric and zinc, the issuing gas will be found to blacken lead paper.

$$H_2SO_3 + 3H_2 = H_2S + 3H_2O.$$

Sulphurous acid with hydrogen yields hydrogen sulphide and water.

A neat method of procuring small quantities of the pure gas is by gently heating a solution of magnesium sulph-hydrate.

$$MgH_2S_2 + 2H_2O = MgO_2H_2 + 2H_2S.$$

The magnesium sulph-hydrate in solution is obtained from alkali waste and sold as a commercial preparation.

Hydrogen sulphide is a colourless gas, with a very unpleasant odour like that of rotten eggs, and poisonous if breathed in quantity. In the air it burns forming water and sulphur di-oxide.

$$H_2S + O_3 = H_2O + SO_2.$$

It can be liquefied by a pressure of seventeen atmospheres. Water dissolves about three volumes of the gas at 15°C., and the solution is slightly acid.

This body is largely used as a reagent in analysis for the separation of metals into groups and the identification of various metals. This may be shown thus:— a series of glasses is arranged containing the following dilute metallic solutions: (1) copper sulphate, (2) cadmium sulphate, (3) antimony chloride, (4) zinc sulphate, containing a little ammonia. On the addition of aqueous solution of hydrogen sulphide to each we obtain metallic sulphides as precipitates, e.g. copper sulphide, brownish black; cadmium sulphide, yellow; antimony sulphide, orange; zinc sulphide, white. In the systematic course of analysis for metals their behaviour with hydrogen sulphide is of great importance, and serves both to separate and identify them.

Vapour density $= 17$ or $H_2S : H_2 = 34 : 2$.

Two volumes of this gas contain two volumes of hydrogen: if therefore the gas is heated in contact with a metal (e.g. tin) which unites with the sulphur no change of volume occurs.

CHAPTER VIII.

THE HALOGEN ELEMENTS.

THE elements **Chlorine, Bromine,** and **Iodine**, form a well-marked natural group, the free elements resembling each other in their general chemical properties and behaviour, and there being also a marked similarity in most of their compounds. They are found widely distributed in nature; in sea water, river water, spring water, and all natural waters, chlorine is invariably found; and in most waters also traces of varying quantities of bromine and iodine may be detected.

Chlorine is obtained, directly or indirectly, from common salt, which is abundant in sea water, and is found in considerable deposits in the form of rock salt. A certain quantity of bromine is obtained from the bitter residual liquors of sea water (containing magnesium bromide) which are left after evaporation and removal of the common salt. The ash of burnt seaweeds (called **Kelp**), furnishes quantities of bromine and iodine; compounds of these elements being withdrawn from sea-water by the living plants and stored in their tissues chiefly in the form of sodium and potassium salts.

Hydrogen Chloride and Hydrochloric Acid. HCl.
Molecular Weight, 36.5. Density, 18.25.

To prepare this compound, a quantity of common salt is placed in a flask, and sulphuric acid passed in by the thistle funnel. The strong acid slightly diluted with water, two of acid to one of water, should be used, and the evolution of the gas assisted if needful by a gentle heat. The gas can be collected in jars either by displacement of air or over mercury, but by reason of its great solubility cannot be collected over water.

The decomposition is thus expressed :—

$$NaCl + H_2SO_4 = NaHSO_4 + HCl.$$

Sodium chloride and sulphuric acid form sodium bi-sulphate and hydrogen chloride. At a high temperature a further decomposition is possible, but is only complete at a red heat. Hydrogen chloride is made on a great scale as a by-product of the alkali manufacture in the preparation of sodium sulphate. The second stage of the reaction is :—

$$NaCl + NaHSO_4 = Na_2SO_4 + HCl.$$

Sodium chloride with sodium bi-sulphate yields neutral sodium sulphate and hydrogen chloride. It may be remarked that many other chlorides treated with sulphuric acid are converted into sulphates with a liberation of hydrogen chloride.

If a solution of the gas is required it may be passed through a series of bottles containing water. Small quantities of the gas are conveniently made by heating the solution: strong hydrochloric acid is placed in a flask (fig. 37) and a gentle heat applied; the evolved gas is dried by passing through strong sulphuric acid, and collected in a tube standing over mercury.

Hydrogen chloride can also be produced by other means, for example, by the direct union of the elements hydrogen and chlorine—

$$H_2 + Cl_2 = 2HCl.$$

Two equal-sized jars are filled with the gases, one

with hydrogen and the other with chlorine, each being closed with a glass plate, and placed together mouth to mouth, the jar containing hydrogen being uppermost; then by slipping away the two plates and inverting the jars the gases can be thoroughly mixed. On applying a lighted taper the elements combine with a sharp explosion, and a cloud of vapour of hydrochloric acid is seen. Exposure to daylight alone will bring about combination, and in sunshine such a mixture instantly explodes.

An important general reaction by which hydrogen

Fig. 37.

chloride is produced, is the decomposition of acid chlorides, such as phosphorus chlorides by water,—

$$PCl_3 + 3H_2O = P(OH)_3 + 3HCl.$$

Phosphorus trichloride with water forms phosphorous acid and hydrogen chloride,—

$$PCl_5 + 4H_2O = PO(OH)_3 + 5HCl.$$

Phosphorus pentachloride with water forms phosphoric acid and hydrogen chloride.

The amount of heat produced by the union of 1 gramme of hydrogen with 35.5 g. of chlorine is very large, viz. 22000 c., and a jet of hydrogen will burn in chlorine as readily as in air, and *vice versa*, chlorine burns readily in hydrogen.

Hydrogen chloride is a colourless and transparent gas, under ordinary conditions dissolving in water, and combining with it with a large development of heat. The solution formed is the powerful acid—**hydrochloric acid**: an older term for it is muriatic acid, or spirit of salt.

The extreme solubility of the gas is shown by having a tube 3 feet long which has been filled by passing a stream of gas for some time to expel all air and then sealed at both ends. On breaking one of the sealed ends under water in a glass, the water rushes up absorbing the gas, and fills the tube almost instantly. Or a bottle filled with the gas and fitted with a tube can be used (fig. 38).

One volume of water dissolves nearly 500 volumes of the gas. The strongest acid has a density of 1·2 and contains 40 per cent. of HCl. Strong hydrochloric acid fumes in damp air, combining with the invisible vapour of water; no visible fumes are formed in dry air. It is a powerful solvent, converting many metals into chlorides with evolution of hydrogen. Its actions with hydrates, carbonates, and peroxides, are shown by the equations:—

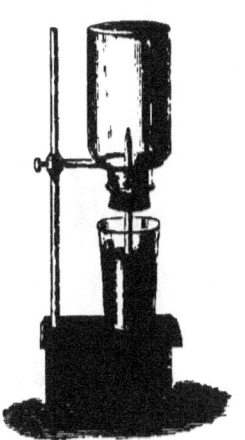

Fig. 38.

$$KHO + HCl = KCl + H_2O.$$
$$CaCO_3 + 2HCl = CaCl_2 + H_2O + CO_2.$$
$$MnO_2 + 4HCl = MnCl_2 + Cl_2 + 2H_2O.$$

Potassium hydrate is converted into potassium chloride, calcium carbonate into calcium chloride, and manganese peroxide into manganese chloride with free chlorine.

The composition of hydrogen chloride by volume is shown by the following experiments.

A U-shaped eudiometer is partly filled with the gas, and the mercury adjusted so as to stand at the same level in both tubes; the gas is then at atmospheric pressure, and its volume can be marked on the tube (fig. 39). The open limb of the tube is then filled with an amalgam of mercury, in which a little metallic sodium has been dissolved, and securely corked, or closed by the thumb, and the gas brought into and shaken with the amalgam. It is afterwards transferred to the closed limb of the tube, and brought to the original atmospheric pressure by running out a little mercury until the remainder stands at the same level on both sides. The gas is decomposed by the sodium amalgam according to the equation:—

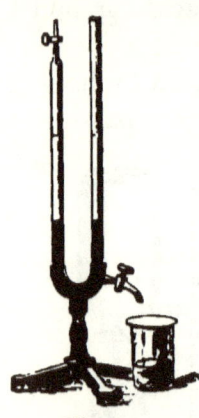

Fig. 39.

$$2HCl + Na_2 = 2NaCl + H_2.$$

Hydrogen chloride and sodium form sodium chloride and hydrogen, and the volume of the hydrogen left is one-half of the volume of the compound gas used. It can easily be shown that the residual gas is hydrogen by expelling it from the stopcock and setting it on fire.

Strong hydrochloric acid is decomposed by an electric current and a mixture of hydrogen and chlorine evolved. We can show that the mixture contains **equal volumes** of chlorine and hydrogen by filling a glass tube with the mixed gases, and allowing the chlorine to be absorbed by potash or potassium iodide. It will be found that one-half the volume (chlorine) is absorbed, and that the residue is hydrogen.

In order to show that the gases unite without change of volume, a glass tube fitted with a stopcock at each end and provided with platinum wires is used. The tube is filled in the dark with the electrolytic mixture of hydrogen and chlorine by passing a stream of the gases (produced by the electrolysis of hydrochloric acid) until all air has been displaced. The stopcocks are then closed, and the contents remain at atmospheric pressure. If a spark is sent between the wires combination takes place, and by opening one stopcock while the end of the tube dips into mercury, it is seen that no change of volume is caused by the union of the gases; but if the open tube is raised into water poured on the surface of the mercury, the gas entirely dissolves, and the tube becomes filled with water. We thus show (1) that gaseous hydrogen chloride contains one-half its volume of hydrogen; (2) that the mixture of gases obtained by electrolysis contains equal volumes of chlorine and hydrogen; (3) that their combination causes no change of volume.

We may represent these facts thus:—

$$\boxed{H_2} + \boxed{Cl_2} = \boxed{HCl \mid HCl}$$

Direct experiments upon the density of these bodies have shown that—

    22·4 litres of hydrogen   weigh 2 grammes
    22·4 ,,   of chlorine   ,,   71   ,,
    22·4 ,,   of hydrogen chloride   ,,   36·5   ,,

from which we see that the density of hydrogen chloride is 18·25, or the exact mean of its components. And according to Avogadro's hypothesis the molecular weight must be 36·5.

Chlorine. Cl. Atomic weight, 35·5. Density, 35·5. Molecular weight, 71 = Cl_2.

The element chlorine is never found in the free state, but is obtained by the decomposition of some chloride; commonly either hydrochloric acid or sodium chloride.

To prepare the gas from hydrochloric acid the dioxide of manganese is used. This oxide is put into a flask fitted with thistle funnel and delivering tube, and strong

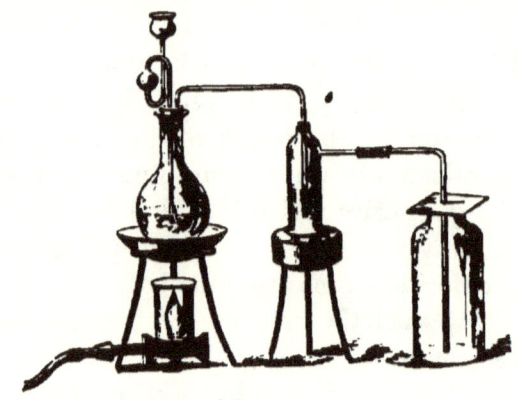

Fig. 40.

hydrochloric acid added; a gradual evolution of chlorine takes place, which can be assisted by very gentle heating. The action is represented as follows:—

$$MnO_2 + 4HCl = MnCl_2 + Cl_2 + 2H_2O.$$

Manganese dioxide with hydrogen chloride yields manganese chloride, chlorine, and water. This is the final stage of the action, a higher chloride of manganese ($MnCl_4$?) being formed as an intermediate product.

The gas should be washed by passing through a little water for the removal of hydrogen chloride, and can be collected by displacement of air (fig. 40), or over *warm* water, but on account of the irritating characters of the gas, it should be made and collected in a cupboard with a good draught.

Or the gas may conveniently be prepared from common salt. A mixture of salt with manganese dioxide is put in the flask, sulphuric acid (previously diluted with its own volume of water) is added, and a gentle heat applied. The equation for the reaction is :—

$$2NaCl + MnO_2 + 3H_2SO_4 = 2NaHSO_4 + MnSO_4 + 2H_2O + Cl_2,$$

the products being sodium sulphate, manganese sulphate, water, and chlorine. In this process the sulphuric acid first causes hydrogen chloride to be liberated, which then reacts with the manganese oxide. Many peroxides, and substances rich in oxygen such as chlorates, nitrates, chromates, etc., cause chlorine to be liberated from hydrogen chloride by converting its hydrogen into water. Their action may be expressed generally thus :—

$$2HCl + [O] = H_2O + Cl_2.$$

At a very high temperature free oxygen is able to liberate chlorine, and, in fact, the gas is produced commercially by passing hydrogen chloride mixed with air over red-hot bricks, impregnated with copper salts.

Chlorine is obtained by the electrolysis of hydrogen chloride, and some metallic chlorides, such as the chlorides of copper, platinum and gold, liberate chlorine when strongly heated.

Chlorine is a yellowish-green gas with an unpleasant odour, causing great irritation of the throat and lungs when breathed. All experiments with this gas, therefore, need caution, and it should not be allowed to escape into the air of the laboratory. It is heavier than air, and soluble in water to the extent of 2 to 2·5 volumes of gas in one volume of water: if the gas be passed into ice-cold water, crystals of chlorine hydrate are formed, $Cl_2, 10\ H_2O$. If these crystals are sealed up in a strong glass tube and placed in warm water, they decompose, and chlorine liquefied by pressure separates from the watery solution as a yellowish liquid.

Chlorine is chemically an extremely active substance, and unites with almost all other elements. The following experiments can be performed to illustrate its properties.

Let a burning taper be plunged into the gas, it is not extinguished, but burns with a smoky flame. The hydrogen of the wax, having a great attraction for chlorine, unites with it, while the carbon is set free as soot.

Turpentine ($C_{10}H_{16}$) is energetically attacked, and a piece of filter paper soaked in that liquid, if brought into a jar of the gas, is frequently set on fire, with formation of fumes of hydrogen chloride and a separation of carbon.

Finely divided metals—**antimony** in powder, or thin **copper** leaf, known as Dutch metal, will burn spontaneously if thrown into the gas.

A piece of **phosphorus** in a deflagrating spoon ignited in air will continue to burn freely if brought into a jar of chlorine, giving off fumes of chlorides of phosphorus:—PCl_3 and PCl_5.

Chlorine gas has great **bleaching** power, and will destroy most vegetable dyes and colours, which may be shown by putting a piece of damp red cotton cloth into a jar of the gas. The removal of the colour is due to an oxidising action, and only takes place in presence of moisture:—

$$H_2O + Cl_2 = 2HCl + [O].$$

Water and chlorine produce hydrogen chloride and oxygen; the latter, however, is not liberated, but enters into combination with the dye in the cloth. Many similar oxidations are effected by chlorine in the presence of water.

Bleaching powder. This substance is a convenient form for utilising chlorine. It is prepared by bringing chlorine gas into contact with slaked lime, with which it forms a compound, $CaO\,Cl_2$, popularly called 'chloride of lime.' Its usefulness depends upon the fact that when treated with dilute acids it evolves chlorine; and very large quantities are used for bleaching, and also for disinfecting purposes.

Bromine. Br. Atomic weight, 80. Density, 80. Molecular weight, 160 = Br_2.

Bromine is obtained for the most part from the mother liquors of brine left from the preparation of salt. Both in the preparation of salt from sea water and from rock salt a quantity of bromides remain in the mother liquors, accompanied with smaller proportions of iodides.

To obtain the bromine the liquors are mixed with **sulphuric acid** and a limited quantity of **manganese dioxide**, the chemical reaction being similar to that in the preparation of chlorine—

$$2NaBr + 3H_2SO_4 + MnO_2 = 2NaHSO_4 + MnSO_4 + 2H_2O + Br_2.$$

The bromine thus set free is distilled over by heating the vessels with steam, and is condensed in large jars. Any iodine present is also set free; but to prevent liberation of chlorine, the manganese dioxide is used in just sufficient quantity to act upon the bromides, and any chlorine set free decomposes a fresh portion of bromide setting an equivalent amount of bromine free.

Bromine is a dark-brown, heavy liquid; its specific gravity is about 3; it becomes solid at −22°C., and boils at 63°. Like chlorine it has an unpleasant odour, and is extremely irritating to the lungs and throat. It dissolves in water to the extent of about 3 parts in 100 of water, forming a reddish liquid. Bromine has some bleaching power, but much less than chlorine.

Hydrogen bromide. HBr. Molecular weight, 81. Vapour density, 40·5.

Hydrogen bromide may be obtained by methods similar to those used in preparing the chloride, but not, however, with the same facility, since this compound is less stable in character. The elements unite with difficulty, but if a mixture of hydrogen with bromine vapour is burnt, or passed over platinum in a red-hot tube, combination takes place.

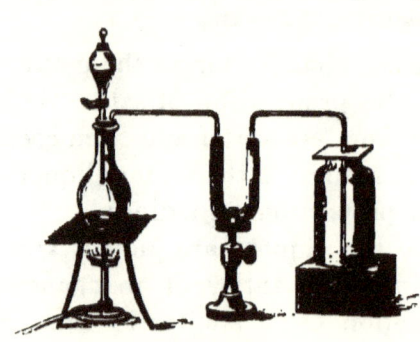

Fig. 41.

When a bromide, such as potassium bromide, is treated with strong sulphuric acid, gaseous hydrogen bromide is liberated, but is partly decomposed, and some free bromine is also produced, by which the gas is coloured brown. Phosphoric acid, however, can be employed instead of sulphuric with success.

The gas is, however, usually made from phosphorus bromide by the action of water.

$$4H_2O + P + Br_3 = 5HBr + H_3PO_4.$$

Amorphous phosphorus is placed in the flask with

a small quantity of water, and bromine gradually added by a tap funnel (fig. 41); the action is very energetic, and the gas comes off freely. To purify it from free bromine it is passed through a tube containing moist amorphous phosphorus, and it may be collected by downward displacement, or over mercury.

Hydrogen bromide is a heavy, colourless gas, fuming in moist air, and dissolving in water to form a strong acid—hydrobromic acid,—which in many respects reresembles hydrochloric acid. When neutralised with alkaline hydrates or carbonates it forms bromides.

$$K_2CO_3 + 2HBr = 2KBr + H_2O + CO_2.$$

Potassium carbonate and hydrobromic acid form potassium bromide, water, and carbon dioxide.

Hydrogen bromide contains equal volumes of hydrogen and bromine united without contraction, and its composition is proved by similar means to those used for hydrogen chloride.

Iodine. I. Atomic weight, 127.
Molecular weight, 254 = I_2 (at low temperatures).

The chief source of iodine is *kelp*, which is the ash obtained by burning seaweeds. Since the kelp contains less than one-half per cent. of iodides, it is subjected to a process of crystallisation to remove the great mass of carbonate, sulphate, chloride, etc., until a mother liquor is obtained containing the iodides and bromides.

The liquor is acidified with sulphuric acid, and manganese dioxide added in small quantity from time to time. Iodine alone is first set free and distilled off, but afterwards more manganese dioxide is added and the bromine comes over. The chemical changes are similar to those with chlorides and bromides.

$$2NaI + MnO_2 + 3H_2SO_4 = 2NaHSO_4 + MnSO_4 + 2H_2O + I_2.$$

Iodine is a gray solid with an almost metallic lustre, and with a peculiar smell somewhat resembling chlorine. It volatilises at ordinary temperatures, and when heated is readily converted into a **violet**-coloured vapour. The specific gravity of the solid is 4·95, nearly five times the density of water; it melts at 114°, and boils at about 200°.

It is slightly soluble in water, readily in alcohol and solution of potassium iodide, producing **brown** solutions; also it dissolves readily in benzene and similar hydrocarbons, in chloroform, and carbon disulphide, and these solutions are **violet** in colour.

Iodine is liberated from most of its compounds with the metals, by chlorine and bromine, but is not displaced from the oxygenated compounds, such as iodates. Free iodine even in minute quantity is recognised by the formation of a **blue** colour with starch solution.

Molecular weight of Iodine. When the vapour density of iodine is determined at temperatures below 590°, it is found to be 127 times as great as that of hydrogen, but at temperatures above this a lower value is found, and the density diminishes to two-thirds of 127, viz. to 85 ($H=1$). Under the reduced pressure of one-tenth of an atmosphere, and at a temperature of 1,300° to 1,400°C., the density falls to 66·3. Now from the first of these results the weight of the iodine molecule becomes $254 = I_2$; but the second value gives 170 as the molecular weight, and the last 132·6. It would seem, therefore, that below 590° the molecule of iodine vapour is I_2, but as the temperature rises a splitting-up of molecules into separate atoms takes place, until the molecular weight finally approaches 127, which is the atomic weight; and probably at higher temperatures the breaking-up would be complete.

$$\boxed{I_2} \quad \text{becomes} \quad \boxed{I} \quad \boxed{I}$$

The molecular weight of bromine is found to be 160 = Br_2 at low temperatures, and to become diminished by about one-sixth at the higher temperatures, which appear to indicate the beginning of a similar decomposition. No change in the density of chlorine has been observed at the highest temperatures.

Hydrogen Iodide. HI. Molecular weight, 128. Density, 64.

The direct union of iodine with hydrogen can be brought about by passing the mixture over heated spongy platinum, when a fuming gas is produced resembling the similar compound of hydrogen with chlorine and bromine.

Hydrogen iodide is also liberated, but mixed with free iodine, when an iodide is decomposed by sulphuric acid.

The method used for preparing the gas is the same as that used for making hydrogen bromide, viz. the decomposition of iodide of phosphorus by water.

$$PI_3 + 3H_2O = P(OH)_3 + 3HI.$$

Phosphorus tri-iodide with water gives phosphorus acid and hydrogen iodide.

If half a gramme of phosphorus, one cubic centimetre of water, and six grammes of iodine are cautiously brought together in a test tube, a very strong solution of hydrogen iodide is formed, which will evolve the gas on heating, and by fixing a bent tube fitted with a cork into the test tube, two or three bottles can be filled with samples of the gas.

Hydrogen iodide is a colourless, very heavy gas, more

than four times as heavy as air, and can be collected by downward displacement. It fumes strongly in air, and is very soluble in water, forming hydriodic acid: both gas and solution are readily decomposed. A little chlorine gas instantly decomposes the gas, producing violet fumes; and also plunging a stout red-hot wire into it liberates violet vapours of iodine. It contains equal volumes of hydrogen and iodine, combined without contraction.

General relations of the halogen elements.

A comparison of the physical and chemical characters of these elements shows the general resemblance of their characters. They exist in the states of gas, liquid and solid respectively, and in boiling-point, specific gravity, atomic weights, and vapour density manifest a regular succession to each other.

The heats evolved by their union with hydrogen are –

$$H + Cl \quad 22,000 \text{ heat units.}$$
$$H + Br \quad 8,440 \quad ,, \quad ,,$$
$$H + I \quad -6,036 \quad ,, \quad ,,$$

The numbers will throw light upon the relative stability of these compounds, since the large evolution of heat in the production of the chloride signifies combination with energy, great chemical attraction, and corresponding stability in the compound; while the negative value of the formation of the iodide means an absorption of energy in its formation, and suggests that it may easily be decomposed into its elements.

Hydrogen bromide or hydrogen iodide are decomposed by chlorine, with a formation of hydrogen chloride and liberation of bromine or iodine; similarly bromine liberates iodine from hydriodic acid; these decompositions are easily understood in the light of the values given above.

CHAPTER IX.

NITROGEN. AMMONIA. NITRIC ACID.

NITROGEN in the free state is one of the most inert of substances, and can only with difficulty be brought into combination; but in the combined form it enters into an infinite number of compounds, organic as well as inorganic, many of which have the most active chemical properties. The compound with hydrogen—ammonia (NH_3) is a powerful alkali, and the oxidised compound nitric acid (HNO_3) is an equally powerful acid.

Ammonia NH_3. Molecular weight 17. Density of gas 8·5.

Ammonia is found in small quantities, in air, earth, and some waters, and in volcanic gases. It has been obtained by the action of an electric discharge upon a mixture of hydrogen and nitrogen.

When animal or vegetable matters, containing nitrogen, putrefy and decay, or are destroyed by heating, ammonia is among the products; and originally an impure ammonia (spirit of hartshorn) was obtained by heating in a retort horn or other similar animal substances. Ammonia in the present day is almost entirely produced by the destructive distillation of coal, as, for example, in the retorts for making illuminating gas, in coke ovens, in blast and other furnaces. The gases given off during the heating of coal are passed through water, which dissolves and retains the ammonia, accompanied by tarry and other substances. From the ammoniacal liquors the ammonia is separated by distillation and led into either hydrochloric or sulphuric acid; thus a salt of ammonia is formed, which may be obtained in the solid state on evaporation of the water, and purified by recrystallisation.

The production of ammonia from animal matter is

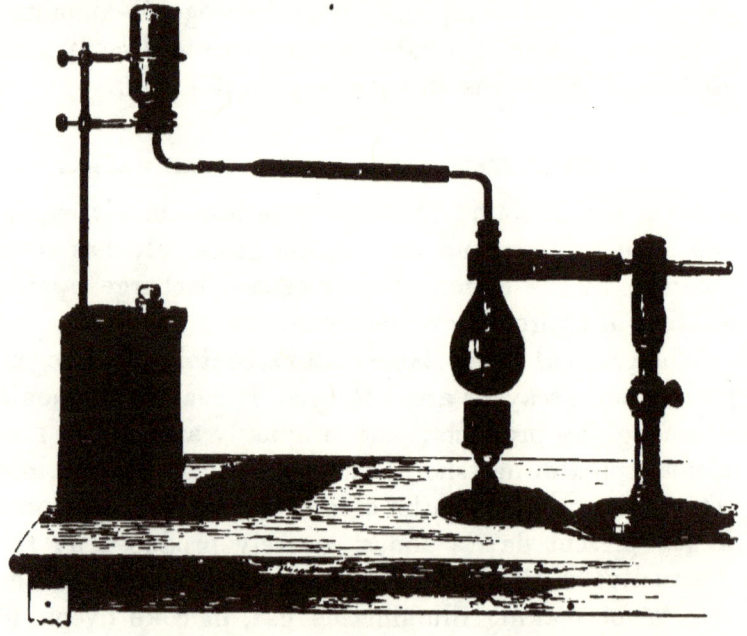

Fig. 42.

readily shown by heating a few pieces of feather in a test tube; the vapours coming off have a strong ammoniacal smell, and will change the colour of red litmus paper to blue, or of yellow turmeric paper to brown, owing to their **alkaline** characters.

Gaseous ammonia is prepared from sal-ammoniac (ammonium chloride) by heating it with lime:—

$$2NH_4Cl + CaO = CaCl_2 + 2NH_3 + H_2O.$$

Ammonia.

Calcium chloride, ammonia, and water are produced. A mixture of the two substances in powder is gently heated in a glass flask or metal vessel; the gas comes off readily, and, being lighter than air, can be collected in bottles held mouth downwards (fig. 42). For drying the gas quick lime is used; sulphuric acid is obviously inadmissible, as it absorbs the gas; and as calcium chloride also forms a compound with ammonia, it cannot be used.

The gas, having a powerful and pungent action on the eyes, nose, and the mucous membranes generally, should not be inhaled, and must be prepared and used with caution.

Ammonia is a transparent, colourless gas, extremely soluble in water: a bottle of gas opened with its mouth under water is filled instantly. Another experiment can be made to show this fact. Let a strong flask or bottle be filled with the gas and the mouth closed by a well-fitted cork, carrying a tube drawn out to a point, at both ends, the lower end being sealed. On bringing the tube into water and breaking the point so rapid an absorption takes place that the water is driven into the bottle with much force (fig. 38).

One gramme of water at $0°$ dissolves nine-tenths of a gramme of ammonia, which is about one thousand-fold its volume. The strong solution of ammonia is known in pharmacy as *liquor ammoniae fortior*, and its density is about ·880 (water = 1). From this solution the gas is freely given off on applying a slight heat, and in this way we may conveniently obtain small quantities for use. If pure ammonia, free from air, is required, the gas should be collected over mercury.

The most important chemical property of ammonia is its **basic** character, i.e. its power to unite with acids and form salts. It has a strongly alkaline taste, turns red litmus paper blue, and turmeric brown. When added to an acid it enters into direct combination to produce a neutral substance or **salt** which does not change the colour of litmus. Thus with hydrogen chloride it forms sal-ammoniac or ammonium chloride:—

$$NH_3 + HCl = NH_4Cl.$$

Take a bottle of gaseous ammonia and another filled with gaseous hydrogen chloride. The bottles should be equal in size, and if the mouths are ground flat they can be covered with greased glass plates to retain the gases; which are slipped aside in making the experiment. The ammonia being lighter should be on the top (fig. 43).

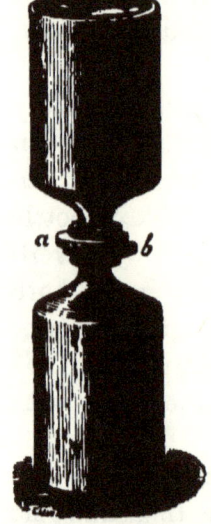

Fig. 43.

When the bottles are placed mouth to mouth, and the glass plates removed, an instant combination takes place, and dense white fumes of ammonium chloride are formed. The 'salt' thus produced, by the combination of the alkaline gas with the acid gas, is a perfectly neutral substance. It has a saline taste, resembling common salt, and it has neither acid nor alkaline properties, and so is without action on litmus paper.

In a similar way we can form ammonium nitrate by mixing solution of ammonia with nitric acid. Both should be somewhat diluted with water, and the nitric acid poured carefully into the ammonia until the liquid is neutral to test paper. Then by gentle heating most of the water is expelled, and the salt will crystallise if the solution be cooled.

Ammonia gas is only feebly combustible in air, its

Ammonia.

hydrogen burning to water, but in oxygen it burns readily, with a greenish flame. If a stream of oxygen gas is from a holder led into a small flask containing the strong solution of ammonia and a light applied, a vigorous combustion takes place. Or, a little strong ammonia is placed in a flask, in the neck of which is a cord supporting a spiral coil of platinum wire (fig. 44). The wire should hang, nearly touching the liquid. If the wire be removed and heated to redness for an instant in a flame, and then replaced while still warm, it will glow and keep red-hot for some time, the heat being maintained by the slow combustion of the mixture of air and ammonia.

Fig. 44.

The composition of ammonia by volume is shown by the eudiometer. About one-third of the closed limb of a U-tube (fig. 39) is filled with gaseous ammonia, and a stream of electric sparks passed between platinum wires sealed into the glass. The ammonia rapidly increases in volume, and after it ceases to expand and the mercury is brought level in both limbs of the tube, the **volume** will be found to be **doubled**. Allowing a portion of the gas to issue from the stopcock, it will be found that the alkaline reaction has disappeared, and the hydrogen in the mixture can be ignited.

In order to determine the composition of the mixture of nitrogen and hydrogen the gas is transferred to a straight eudiometer and measured. Then pure oxygen is added, in excess and measured, and the mixture exploded; the water formed condenses, and a residue of nitrogen with oxygen is left, which is also measured.

Since two volumes of hydrogen with one volume of oxygen unite to form water, the amount of hydrogen in the mixture will be two-thirds of the contraction on explosion.

Let us suppose, for example, an experiment yields the following measures:—

Volume of Ammonia	= 20 cc.
Mixture after passing the spark	= 40 cc.
After adding oxygen	= 60 cc.
After explosion	= 15 cc.

The contraction is, therefore, 45 cc., two-thirds of which gives hydrogen, equals 30 cc., and by difference the nitrogen equals 10 cc.

$$NH_3 \text{ (2 vols.)} = N \text{ (1 vol.)} + H_3 \text{ (3 vols.)}$$

That is, two vols. of ammonia yield one volume of nitrogen and three vols. of hydrogen. The vapour density of ammonia has been found by direct experiment to be 8.5, which is also in agreement with this result. The molecular weight of ammonia, therefore, is 17.

Nitric Acid. Hydrogen Nitrate HNO_3. Molecular weight 63.

Nitric acid is prepared from an alkaline nitrate: a quantity of potassium nitrate (nitre) is placed in a retort, sulphuric acid is added, and the mixture gently heated. The vapours of nitric acid distil over, and condensing in the neck of the retort are collected in a well-cooled receiver fitted to the extremity (fig. 20). The action is as follows:—

$$KNO_3 + H_2SO_4 = KHSO_4 + HNO_3.$$

Potassium nitrate and sulphuric acid produce acid potassium sulphate and nitric acid.

For the manufacture of the acid on a large scale sodium nitrate (Chili saltpetre) is used, and the opera-

Nitric Acid.

tion is carried on in large iron vessels. Double the quantity of sodium nitrate may be used, but the higher temperature required to finish the action causes some nitric acid to decompose, and the product is more or less coloured yellow by oxides of nitrogen.

$$2\,NaNO_3 + H_2SO_4 = Na_2SO_4 + 2\,HNO_3.$$

Two molecules of sodium nitrate with one of sulphuric acid produce neutral sodium sulphate and nitric acid.

Pure nitric acid is a colourless liquid; its density is above 1·5, but the acid commonly used in the laboratory is of the specific gravity 1·42, and then contains 70 per cent. of real acid, HNO_3, and 30 per cent. of water. Acid of this strength may be distilled without decomposition, but a stronger acid is partly decomposed on distillation. At a red heat nitric acid is decomposed entirely into nitrogen peroxide, oxygen, and water.

$$2\,HNO_3 = 2\,NO_2 + O + H_2O.$$

A slight decomposition takes place by simple exposure to light, the acid acquiring a yellow tinge.

Nitric acid is a strongly corrosive substance and a powerful solvent. It acts energetically on many inorganic and organic bodies; the elements sulphur, phosphorus, and iodine are by it oxidised and converted into sulphuric, phosphoric, and iodic acids; copper, silver, mercury, and other metals are dissolved and converted into nitrates, but gold and platinum are not attacked even by the most concentrated acid.

Many organic substances are acted upon by nitric acid yielding nitro-derivatives, some of which are powerful explosives, for instance, nitro-glycerine and gun-cotton. The formation of nitro-benzene will serve as an example of the reaction.

$$C_6H_6 + HNO_3 = C_6H_5, NO_2 + H_2O.$$

Benzene and nitric acid form nitro-benzene and water.

Nitric acid mixed with hydrochloric acid forms the liquid termed *aqua regia*, a most powerful solvent; it dissolves most metals, including gold and platinum, forming metallic *chlorides*.

Nitric acid can be produced by *synthesis*, starting either from nitrogen or ammonia:—

(1) When a series of electric sparks are passed through a mixture of nitrogen and oxygen, contained in a globe, red fumes are formed; which in contact with water produce nitric acid. It is probable that traces of nitric acid are formed in this manner in the air during thunderstorms.

(2) Let a jet of hydrogen be burnt in air so that the water formed can be collected; it will be acid and will contain nitric acid.

(3) The oxidation of ammonia produces nitrite and nitrate. This oxidation may be effected by burning the ammonia in air or oxygen; by passing ammonia with excess of air over heated spongy platinum; or by the action of ozone on very dilute ammonia.

$$2NH_3 + O_3 = NH_4NO_2 + H_2O.$$
$$2NH_3 + O_4 = NH_4NO_3 + H_2O.$$

(4) **Nitrification.** Nitrates are found in soils in all parts of the world, being continually formed by the oxidation of ammonia and organic substances, and plants depend largely for the nitrogen they require upon the nitrates thus produced. In the neighbourhood of stables, manure heaps, cesspools, and similar places nitrates are abundant, and occasionally potassium nitrate is found as an efflorescence (saltpetre) upon dry soil. In this country calcium nitrate is more commonly present in soils, but does not appear as an efflorescence, as it is very soluble and deliquescent. A quantity of sodium

nitrate, or *Chili saltpetre*, is brought from rainless districts in South America, where large deposits of unknown origin exist.

Artificial nitre beds (nitre plantations) have been constructed for the production of this salt on the large scale. A suitable porous soil mixed with animal refuse and wood ashes (the latter to furnish potassium) is watered from time to time with animal liquids or sewage until an accumulation of nitre takes place. The soil is then washed with water to dissolve out the nitrates which are recovered from solution after evaporation and crystallising.

Although the actual production of nitrates from ammonia in soils is effected by the process of oxidation by the oxygen of the air, it is found that 'nitrification' depends upon the presence of minute organisms, 'micrococcus nitrificans,' and if a soil be sterilised by heating or other means it loses the power of causing this change.

Action of Metals on Nitric Acid.

When hydrochloric acid or dilute sulphuric acid acts upon zinc, a salt is formed and hydrogen set free. In these cases there appears to be a simple replacement of hydrogen by metal.

$$H_2SO_4 + Zn = ZnSO_4 + H_2.$$

But when copper acts upon strong sulphuric acid no hydrogen is obtained, but sulphur dioxide is produced by a further decomposition of the acid. This production of sulphur dioxide appears to arise from two successive changes, the first being represented thus:—

$$Cu + H_2SO_4 = CuSO_4 + [H_2],$$

the hydrogen sulphate becoming copper sulphate, but the hydrogen, instead of being liberated as gas, attacks the sulphuric acid and deprives it of part of the oxygen, thereby producing sulphur dioxide and water:—

$$H_2SO_4 + H_2 = SO_2 + 2H_2O.$$

And the reducing action of the hydrogen may proceed still further until free sulphur, and even hydrogen sulphide are obtained. The black residue left after the action of copper on sulphuric acid contains copper sulphide.

Nitric acid with metals is acted upon in a similar manner, no hydrogen being liberated, but a series of products obtained by successive removals of oxygen.

When copper dissolves in nitric acid of sp. gr. 1·2, the gas produced contains 98 per cent. of nitric oxide (NO), 1 per cent. of nitrous oxide (N_2O), and 1 per cent. of nitrogen (N_2). But if the acid be diluted with water (one part of acid to eight of water), as much as 20 per cent. of nitrous acid (N_2O) and 7 per cent. of nitrogen are obtained.

Again, zinc with the same dilute acid gives a mixture of gases containing 46 per cent. of nitric oxide (NO), 50 per cent. of nitrous oxide (N_2O), and 4 per cent. of nitrogen; besides which a considerable amount of ammonia as nitrate is left in the solution.

The strength of the acid, the temperature, and other conditions affect the quantities of the resulting products, the most complete reduction being obtained by prolonged action with dilute acids. We may represent the reductions by the following equations:—

$$HNO_3 + H_2 = HNO_2 + H_2O. \quad \text{Nitrous acid.}$$
$$HNO_2 + H_2 = HNO + H_2O. \quad \text{Hyponitrous acid.}$$
$$HNO + H_2 = H_3NO. \quad \text{Hydroxylamine.}$$
$$H_3NO + H_2 = H_3N + H_2O. \quad \text{Ammonia.}$$

But as these compounds cannot exist in the presence

of each other and of nitric acid, a series of secondary changes occur, resulting in the liberation of nitrogen oxides and also free nitrogen :—

$$3HNO_2 = HNO_3 + 2NO + H_2O.$$
$$HNO_3 + NH_3 = N_2O + 2H_2O.$$
$$HNO_2 + NH_3 = N_2 + 2H_2O.$$

CHAPTER X.

OXIDES OF NITROGEN.

The series of oxides of nitrogen is complete from the lowest (N_2O) to the highest (N_2O_5) :—

N_2O_5 Nitrogen pentoxide or Nitric anhydride.
N_2O_4 or NO_2 Nitrogen tetroxide or Nitrogen peroxide.
N_2O_3 Nitrogen trioxide or Nitrous anhydride.
$[N_2O_2]$ or NO Nitrogen dioxide or Nitric oxide.
N_2O Nitrogen monoxide or Nitrous oxide.

Nitrogen Pentoxide or Nitric Anhydride N_2O_5.

This substance is produced from nitric acid by the removal of water; it is a solid, unstable substance, and combines eagerly with water reproducing nitric acid.

One method for preparing it is to act upon the most concentrated nitric acid with phosphorus pentoxide.

$$2HNO_3 + P_2O_5 + = 2HPO_3 + N_2O_5.$$

Nitric acid with phosphorus pentoxide forms phosphoric acid and nitrogen pentoxide. The reproduction of nitric acid is expressed by the equation :—

$$N_2O_5 + H_2O = 2HNO_3.$$

Nitrogen peroxide or Nitrogen tetroxide N_2O_4 or NO_2.

This oxide is obtained as a reddish gas when nitric oxide is mixed with excess of oxygen or air :—

$$2NO + O_2 = 2NO_2.$$

It is also given off when lead nitrate is heated :—

$$PbN_2O_6 = PbO + N_2O_4 + O.$$

The products of the decomposition are lead oxide, nitrogen peroxide, and oxygen.

If lead nitrate be heated in a retort and the red vapours produced are led into a bent tube immersed in a freezing mixture, a reddish brown liquid is condensed and the oxygen passes on (fig. 45). This liquefied oxide boils at 22°C., and may be kept in well-closed vessels.

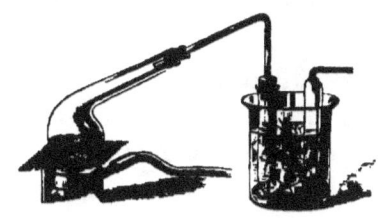

Fig. 45.

It has no acid properties when dry, but with water it is split up in one of two ways:—

(1) $\quad N_2O_4 + H_2O = HNO_3 + HNO_2,$

With a little water it produces **nitric and nitrous acids**;

(2) $\quad 3N_2O_4 + 2H_2O = 4HNO_3 + 2NO,$

with excess of water (owing to the unstable character of nitrous acid) the products are nitric acid and nitric oxide.

This oxide contains 14 parts of nitrogen with 32 of oxygen, or nearly 70 per cent. of oxygen; and on account of the large quantity of oxygen present, and also the ease with which it breaks up, the liquid oxide has been used as an ingredient for making explosives.

The **vapour density** of this oxide, and in consequence also its molecular weight, varies with the temperature. It is found that at low temperatures, 26°C., the density is 41·4, which diminishes quickly as the temperature rises, until at 140°C. the density is 23.

The formula N_2O_4 corresponds to a molecular weight 92 and density 46; while NO_2 expresses the molecular weight 46 and density 23. We see, therefore, that at high temperatures the gas consists entirely of molecules of NO_2, but at low temperatures it is probably a mixture of molecules of N_2O_4 and NO_2. At 26° the composition would be 80 per cent. and 20 per cent. respectively.

A visible effect of this molecular change is the change in the colour of the gas produced by heating. If a thin flask filled with the gas is gently heated, the pale-red gas gradually darkens and becomes more opaque until it appears almost black in colour.

[At temperatures approaching a red heat a further change takes place, and the gas decomposes into nitric oxide and oxygen.]

Nitrogen trioxide or Nitrous anhydride. N_2O_3.

The trioxide is the most unstable of the oxides of nitrogen. It is prepared from nitric acid by heating with starch or arsenic trioxide, which become oxidized in the operation.

$$HNO_3 + As_2O_3 + H_2O = H_3AsO_4 + N_2O_3.$$

Arsenic trioxide and nitric acid produce arsenic acid and nitrogen trioxide.

The red fumes given off are a mixture of dioxide with tetroxide, but if passed into a freezing mixture a green liquid is condensed which is chiefly trioxide. It is very unstable, and if the temperature be allowed to rise decomposes again into dioxide and tetroxide.

$$2N_2O_3 = N_2O_4 + 2NO.$$

With ice-cold water nitrogen trioxide forms nitrous acid, also an unstable substance, for it breaks up if warmed into nitric acid and nitrogen dioxide:—

$$N_2O_3 + H_2O = 2HNO_2.$$
$$3HNO_2 = HNO_3 + 2NO + H_2O.$$

Nitrites. Although nitrous acid cannot be isolated as a definite compound, it forms by uniting with metals perfectly stable and definite salts; such as potassium nitrite, KNO_2, sodium nitrite, $NaNO_2$, silver nitrite, $AgNO_2$.

If the red fumes given off by heating together starch and nitric acid are passed into solution of potash or soda the nitrite is produced.

Potassium nitrite is also produced by strongly heating potassium nitrate which decomposes with loss of oxygen:—

$$KNO_3 - O = KNO_2.$$

Nitrogen dioxide or Nitric Oxide, NO. Molecular weight 30, Density 15.

This oxide is most readily produced from nitric acid by the action of metallic copper.

Pieces of sheet copper are placed in a flask, fitted with thistle funnel and delivery tube (fig. 1), and nitric acid, previously mixed with an equal volume of water, poured upon the metal. A very gentle heat may be necessary to start the action; the gas which comes off can be collected over the water.

The equation for the reaction is—

$$3Cu + 8HNO_3 = 3CuN_2O_6 + 4H_2O + 2NO.$$

Copper and nitric acid produce copper nitrate, water, and nitric oxide. The gas is not pure but contains nitrous oxide as well as free nitrogen.

Nitric oxide is a colourless gas, almost as difficult to liquefy as oxygen or nitrogen. It is neutral to test-paper, but on exposure to air combines with free oxygen, forming red fumes which dissolve in water with an acid reaction. The density of the gas is 15, and it contains equal volumes of nitrogen and oxygen.

This is proved by the observation that if iron or tin be strongly heated in the gas, an oxide is formed and the residue of nitrogen left is one half the volume of gas originally taken.

The following experiments may be made with samples of the gas :—

(1) A jar is half filled with oxygen in the pneumatic trough and nitric oxide added little by little. As the gases mix red fumes form (N_2O_4), but soon dissolve in the water forming nitric and nitrous acids.

(2) A similar experiment can be performed with air. A tube is taken graduated or divided into ten equal parts: *five volumes* of air are passed in, and then by degrees two volumes of nitric oxide added. The free oxygen uniting with nitric oxide, forms the peroxide which dissolves in the water; and finally *four volumes* of nitrogen are left. The experiment thus serves as a rough method for the analysis of air and was used by Cavendish for that purpose.

(3) Nitric oxide will support the combination of many bodies if the temperature be sufficiently high to decompose the gas. Thus if the gas be tested with a taper just lit, the taper will not burn, but a taper burning briskly will burn more brightly still if the gas contain a little monoxide (N_2O). Sulphur and phosphorus if plunged into a jar of nitric oxide while feebly burning are extinguished, but in full combustion the latter will burn almost as brightly as in oxygen.

(4) Pour a few drops of carbon disulphide into a jar of the gas and shake the mixture. If a light is applied to the mouth of the jar, a dazzling flash of flame is produced.

(5) Pour a little solution of iron protosulphate into a jar of the gas; the nitric oxide will dissolve forming a dark coloured solution: this fact is used as a test for nitrates as follows :—

A few fragments of nitre are placed in a test-tube and moistened with water, a little strong sulphuric acid is added, and a solution of ferrous sulphate slowly poured into the test-tube, which is held in a slanting position. If the iron solution is carefully added it will float on the top of the denser acid, and at the line of junction a **brown ring** will form.

The chemical change is thus represented:—

$$6FeSO_4 + 5H_2SO_4 + 2KNO_3 = 3Fe_23SO_4 + 2KHSO_4 + 2NO + 4H_2O.$$

Ferrous sulphate, sulphuric acid, and potassium nitrate, form ferric sulphate, potassium sulphate, nitric oxide, and water. The nitric oxide dissolving in the iron solution produces the brown colour.

Nitrogen Protoxide. Nitrous Oxide. Laughing Gas. N_2O. Molecular weight 44. Density 22.

This gas is usually prepared by heating ammonium nitrate, but may also be obtained by the reduction of nitric acid, by zinc or other metals.

Thirty or forty grammes of dry ammonium nitrate are placed in a retort and gradually heated. The salt first fuses, and soon begins to decompose, giving off bubbles of gas; a gentle heat only should be applied, or the decomposition may take place too rapidly and even explosively. The equation for the change is—

$$NH_4NO_3 = N_2O + 2H_2O.$$

Steam and nitrous oxide being the products.

The gas can be collected over warm water; in which it is somewhat soluble, but to a less extent than in cold water.

Nitrous oxide is a colourless gas with a slightly sweet taste and smell. If breathed in small quantity mixed with air it produces a transient intoxication, whence its name 'laughing gas,' but the pure gas when inhaled quickly produces complete insensibility. On account of this property it is used, especially by dentists, as an anæsthetic. The gas does not affect litmus, having no acid properties; it does not form red fumes with oxygen or air, thus differing from nitric oxide. It is a good supporter of combustion, but the combustible substance needs to be burning briskly when brought into the gas, and at a temperature sufficiently high to decompose it, setting free the oxygen.

For example, a glowing splinter of wood kindles and burns brightly in the gas. Let a fragment of sulphur in a spoon be ignited and brought into the gas, it is extinguished; but if strongly heated and again plunged in it burns vividly. A similar experiment can be tried with phosphorus.

The composition of nitrous oxide is shown by exploding a mixture of the gas with hydrogen in the eudiometer; a quantity of nitrogen equal in volume to the original gas is left.

$$N_2O + H_2 = H_2O + N_2.$$

Two volumes of nitrous oxide with two volumes of hydrogen leave two volumes of nitrogen. At a red heat, 900° C., two volumes of nitrous oxide yield two volumes of nitrogen and one volume of oxygen. Its density is 22, and its molecular weight 44.

$$[N_2O = 28 + 16 = 44.]$$

Hyponitrites. A nitrite or nitrate reduced by sodium amalgam yields a solution containing a hyponitrite. The sodium salt NaNO, and the silver salt AgNO are known, but the acid does not exist in a free state.

Hydroxylamine $NH_2.OH$.

This body is obtained in small quantity by the reduction of nitrates, nitrites, and nitric oxide, etc., by the agency of hydrogen, which is generated for this purpose by the action of tin or hydrochloric acid.

It is a slightly basic substance, forming a compound with hydrochloric acid NH_2OH, HCl, and it exerts a reducing action with metallic salts by depriving them of oxygen.

Diamine. N_2H_4. NH_2, NH_2.

This interesting compound has recently been obtained. It is a very stable gas, with a peculiar odour, but scarcely like ammonia; when breathed it powerfully affects the nose and throat. The gas is very soluble in water, turns red litmus paper blue, and the solution reduces silver and copper salts to the metallic state. Diamine is a strongly basic substance and forms well-defined crystalline salts. The sulphate has the composition N_2H_4, H_2SO_4 and the hydrochlorate $N_2H_4, 2HCl$.

CHAPTER XI.

Phosphorus. P. Atomic weight, 31. Molecular weight, $124 = P_4$.

PHOSPHORUS is one of the elements never found free in nature; but in the form of phosphate it is widely distributed and occurs in most soils. The chief minerals containing it are phosphorite, apatite and coprolites. Being a necessary constituent of plants and essential to their growth, it is taken up by them and accumulated chiefly in the seeds. Animals in turn use phosphorus thus stored, and it appears in their blood, brain, and many tissues, and especially in the bones, the earthy portion of which consists very largely of calcium phosphate.

The element phosphorus is manufactured from calcium phosphate—bone ash, or some mineral form of the same substance. The first stage in the operation results in the production of **phosphoric acid.** The calcium phosphate is treated with sulphuric acid in excess whereby most of the calcium is converted into sulphate and solution of phosphoric acid obtained. The change is thus expressed—

$$Ca_3P_2O_8 + 3H_2SO_4 = 3CaSO_4 + 2H_3PO_4.$$

The liquid is filtered from the calcium sulphate, evaporated to a syrup, mixed with charcoal powder, dried, and finally strongly heated to drive off water. The effect of the heat is to convert the ortho-phosphoric acid into meta-phosphoric acid.

$$H_3PO_4 - H_2O = HPO_3.$$

The next stage of the process consists in the reduction of the phosphoric acid to phosphorus by means of carbon.

The mixture of meta-phosphoric acid with carbon is placed in a retort fixed in a furnace and brought to a bright red heat, when the phosphorus is set free, and passing over in vapour is received in water, the gaseous carbonic oxide being allowed to escape. This change is as follows:—

$$2HPO_3 + 5C = P_2 + H_2O + 5CO.$$

The manufacture of phosphorus is a difficult and dangerous operation to carry out, on account of the inflammable character of the element.

The crude phosphorus is remelted and purified, and usually cast into the form of sticks.

Common phosphorus is a waxy solid, soft enough to be cut with a knife, but becoming brittle and crystalline at low temperatures. It inflames with great readiness, and will take fire if left exposed to the air. Burns on the skin made by phosphorus are very painful and difficult to heal, and great caution ought always to be used in making experiments with phosphorus. It should be kept in water, and always cut beneath the surface of water, and should not be handled with the fingers.

At a temperature of 44° phosphorus melts into a transparent liquid with a very slight tinge of yellow: it solidifies to a transparent waxy solid, but after exposure to light the surface becomes coated with an opaque film of **amorphous** phosphorus, which darkens gradually until it becomes almost black. Waxy phosphorus will dissolve in turpentine and to a certain extent in some oils: it is freely soluble in carbon-disulphide, but the solution if exposed to light becomes turbid and deposits *amorphous* phosphorus.

Beautiful **crystals** of phosphorus may be obtained by

slow evaporation in the dark in a vacuous tube. A piece of dry phosphorus is placed in a glass tube, which, after being exhausted of air by a Sprengel pump, is sealed; and if the tube be kept in the dark for some time, a number of brilliant glittering crystals will be found adhering to the sides. The crystals will keep their beauty and transparency unless exposed to light.

The density of ordinary phosphorus is 1·8 : it boils at 290°C., and its vapour density is 62. The molecular weight therefore is $124 = P_4$.

Allotropic Forms.

Ordinary waxy phosphorus readily alters by light or heat into **amorphous** or red phosphorus, which differs from it in many important characters. If a stick of phosphorus sealed in a glass tube is heated to a temperature of about 250°, it changes very gradually into the red variety becoming more opaque as the change proceeds, until finally it is converted into a hard red mass. There is no change of weight in the operation, but the transformation is attended with an evolution of heat. The red variety is even more easily produced by heating phosphorus with a little iodine. In sealed tubes red phosphorus can be heated to a very high temperature, 500°, and sublimed without fusion (changing into a third crystalline variety), but if heated to 260° in contact with air it passes suddenly into the ordinary form and takes fire. The density of red phosphorus is 2·1, and that of the crystalline form is 2·4.

Amorphous phosphorus differs from the ordinary kind in not being soluble in carbon disulphide, and much more difficult to oxidise. It is not poisonous, whereas ordinary phosphorus is an active and violent poison.

Phosphorus is largely used for the manufacture of

lucifer matches. In the older kinds of matches the composition used for tipping contains ordinary phosphorus, but in the so-called 'safety matches' the composition on the tips contains no phosphorus, but a paste containing red phosphorus is on the box, and on this they ignite by gentle friction.

Heat of Combustion. The great energy with which phosphorus burns in oxygen is shown by the large amount of heat produced:—

1 gramme of phosphorus in oxygen gives 5868 c.
i. e. 1 atom of phosphorus in oxygen (or 31 grammes of P. with 40 grammes of O.) gives 181,900 c.
The combustion in chlorine gives the values
$P + Cl_3$ Phosphorous trichloride, 75,800 c.
$P + Cl_5$ Phosphorous pentachloride, 107,800 c.

The principal compounds of phosphorus are shown in the following table:—

P_2O_3, P_2O_4, P_2O_5, oxides.
PCl_3, PCl_5 chlorides.
H_3P, phosphine or phosphuretted hydrogen.
H_3PO, phosphamine? unknown.
H_3PO_2, hypophosphorous acid.
H_3PO_3, phosphorous acid.
H_3PO_4, phosphoric acid.

Oxides of Phosphorus.

Phosphorus burns readily in air or oxygen, and provided sufficient oxygen is supplied forms the pentoxide, P_2O_5. At ordinary temperatures phosphorus oxidizes slowly, appearing luminous in the dark. When, however, the phosphorus is burnt in with a deficiency of oxygen lower oxides P_4O, P_2O_3, P_2O_4 are obtained.

Phosphorus Pentoxide, P_2O_5, is a white powder formed by burning phosphorus in *excess* of air or oxygen. It is unaltered by heat, being neither fusible nor volatile. It combines eagerly with water to form phosphoric acid; and on account of its great affinity for water is a powerful drying agent: and it is especially useful for drying gases when the last traces of moisture have to be absorbed.

$$P_2O_5 + 3H_2O = 2H_3PO_4.$$

Phosphorus pentoxide with water forms phosphoric acid.

Phosphorus trioxide P_2O_3, is obtained as a white powder by burning phosphorus in a glass tube through which air is slowly passed from an aspirator. It is never pure, but contains pentoxide in admixture. With water it forms phosphorous acid.

$$P_2O_3 + 3H_2O = 2H_3PO_3.$$

If the product obtained as above described is heated in a vacuous tube the trioxide appears to decompose into sub-oxide and tetroxide.

$$7P_2O_3 = P_4O + 5P_2O_4.$$

Chlorides of Phosphorus. $PCl_3, PCl_5.$

It has been shown that phosphorus will burn in a jar of chlorine gas. The product of the combustion is

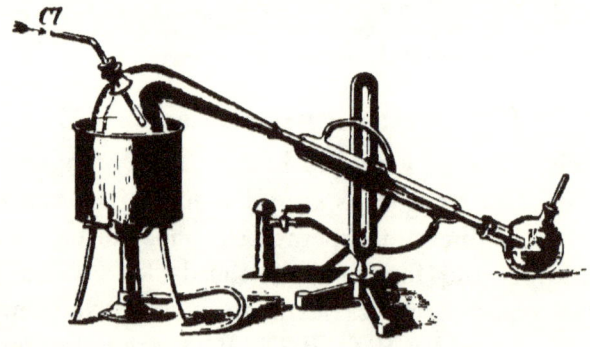

Fig. 46.

pentachloride, PCl_5, when an excess of chlorine is used, or trichloride, PCl_3, with a deficiency of chlorine.

To prepare these compounds, phosphorus is placed in a retort connected with a cooled receiver and a stream of chlorine passed into the apparatus; the retort is bedded in sand and kept heated (fig. 46). The trichloride of phosphorus which is first formed passes on and condenses as a **liquid** in the receiver. When all the phosphorus is converted into trichloride the liquid can be poured back into the retort, and by keeping up the current of chlorine it becomes converted into the solid pentachloride.

(i) $P + Cl_3 = PCl_3$.
(ii) $PCl_3 + Cl_2 = PCl_5$.

The preparation of these bodies is troublesome and unpleasant, on account of the irritating characters of the compounds themselves as well as those of the chlorine employed in their production.

Phosphorus pentachloride (PCl_5) is a crystalline solid of a pale yellow colour. It can be sublimed at 100°, but at a somewhat higher temperature, 150°, begins to break up into trichloride and free chlorine. The *dissociation* is complete at about 340°. It fumes unpleasantly in moist air, giving off clouds of hydrochloric acid, and is very rapidly decomposed by water. With a little water phosphorus oxychloride is formed thus:—

$$PCl_5 + H_2O = PO.Cl_3 + 2HCl,$$

but an excess of water converts it entirely into phosphoric acid.

$$PCl_5 + 4H_2O = PO(OH)_3 + 5HCl.$$

Not only does phosphorus pentachloride react with water in this manner, but similar reactions take place with a large number of other bodies, and this compound is frequently used for the purpose of replacing the group (O H) 'hydroxyl' by chlorine. The following examples will show the action :—

$$SO_2(OH)_2 + PCl_5 = SO_2(OH)Cl + HCl + POCl_3.$$

With sulphuric acid, chlorosulphuric acid is produced together with hydrogen chloride and phosphorus oxychloride.

$$C_2H_5(OH) + PCl_5 = C_2H_5Cl + HCl + POCl_3.$$

Alcohol (ethyl hydrate) is converted into ethyl chloride.

$$(C_2H_3O)OH + PCl_5 = (C_2H_3O)Cl + HCl + POCl_3.$$

Acetic acid yields acetyl chloride, etc.

Phosphorus trichloride, PCl_3. The trichloride is a liquid boiling without decomposition at 76°. It fumes in moist air, and with water decomposes, forming phosphorous acid.

$$PCl_3 + 3H_2O = P(OH)_3 + 3HCl.$$

Its vapour density is 68·75, and molecular weight 137·5 = PCl_3.

The vapour density of the pentachloride is abnormal on account of a dissociation of the compound by heating into trichloride and free chlorine. The theoretical number for the molecule PCl_5 is 104·2, but by experiment the actual density is found to be 52 at 340°, increasing to 72 at 180°. The splitting up is apparently complete at the higher temperature, but partial only at the lower. It has however been found that the decomposition is reduced to a minimum in an excess of chlorine; and determinations of the vapour density of phosphorus pentachloride volatilised in a globe previously filled with chlorine, give values agreeing approximately with the formula PCl_5.

Phosphorus also forms compounds with bromine, iodine and fluorine; PBr_3 PBr_5; PI_3 P_2I_4; PF_5.

Phosphoric Acid. H_3PO_4.

Phosphoric acid may be obtained from phosphorus (1) by burning it and causing the pentoxide to unite with water, or (2) by heating phosphorus with diluted nitric acid until it dissolves, and evaporating the liquid to a syrup. The excess of nitric acid is expelled and the pure acid left.

For commercial purposes, however, phosphoric acid is prepared directly from bone ash by the action of sulphuric acid, as in the first step towards the production of phosphorus. Finely ground bone ash ($Ca_3P_2O_8$) is treated with sulphuric acid; the calcium is separated in part as gypsum (calcium sulphate), and the acid liquid so obtained is concentrated to a small bulk. A further addition of sulphuric acid is made, which precipitates more calcium sulphate; and the acid liquid, after filtration, again concentrated, and finally heated to expel excess of sulphuric acid. If necessary, the treatment with sulphuric is repeated a third time, to remove the last traces of calcium.

A solution of phosphoric acid after evaporation yields a syrupy acid liquid, which after a time will crystallise. This acid is called **orthophosphoric**. By heating, it loses water; two distinct products are obtained; first at 213°—

$$2H_3PO_4 - H_2O = H_4P_2O_7,$$

pyrophosphoric acid is formed, while at a red-heat another molecule of water is lost:—

$$H_4P_2O_7 - H_2O = 2HPO_3.$$

Or

$$H_3PO_4 - H_2O = HPO_3.$$

This compound is called **metaphosphoric acid**, and cannot be further dehydrated by heat.

It is evident that the three acids are compounds of the pentoxide (phosphoric anhydride) with varying amounts of water.

$$P_2O_5 + 3H_2O = 2H_3PO_4 \quad \text{ortho-phosphoric acid.}$$
$$P_2O_5 + 2H_2O = H_4P_2O_7 \quad \text{pyro-phosphoric acid.}$$
$$P_2O_5 + H_2O = 2HPO_3 \quad \text{meta-phosphoric acid.}$$

We have seen that by loss of water one can be produced from the other; so by the addition of water to the lower hydrates, orthophosphoric acid may be produced.

Orthophosphoric acid is a strong acid, with a very sour taste, and in combination with metals forms a large number of salts. With sodium, for example, it forms three compounds, containing one, two, or three atoms of metal to one of phosphorus and three of oxygen in each.

H_3PO_4, ortho-phosphoric acid.
NaH_2PO_4, monosodiumphosphate, or acid phosphate.
Na_2HPO_4, disodium phosphate, or neutral phosphate.
Na_3PO_4, trisodium phosphate, or basic phosphate.

The first compound has an acid reaction; the second is neutral, and is, in fact, the common sodium phosphate used as a reagent in the laboratory; the third is an alkaline body.

Some other phosphates of importance are—

$NaNH_4HPO_4$, sodium ammonium phosphate, microcosmic salt.
$MgNH_4PO_4$, magnesium ammonium phosphate.
Ag_3PO_4, yellow phosphate of silver.
$Ca_3P_2O_8$, tricalcium phosphate.
$CaH_4P_2O_8$, monocalcium phosphate, or 'superphosphate.'

Microcosmic salt is obtained by dividing equivalent quantities of disodium phosphate and ammonium chloride, and crystallising.

$$Na_2HPO_4 + NH_4Cl = NH_4NaHPO_4 + NaCl.$$

Or by mixing diammonium phosphate with disodium phosphate. It decomposes by heating, leaving a glassy mass of sodium metaphosphate.

$$NaNH_4HPO_4 = NaPO_3 + NH_3 + H_2O.$$

Ordinary sodium phosphate (rhombic phosphate) is made by adding caustic soda or sodium carbonate to phosphoric acid from bone ash, until the liquid is slightly alkaline and crystallising. The crystals contain

$Na_2HPO_4, 12H_2O$. When heated to redness, sodium pyrophosphate remains.

$$2Na_2HPO_4 = Na_4P_2O_7 + H_2O.$$

Magnesium ammonium phosphate, $MgNH_4PO_4\ 6H_2O$, is a crystalline precipitate, obtained by adding sodium phosphate to a solution of magnesium salt containing ammonia; the formation of this precipitate is one of the tests for phosphoric acid or for magnesium.

Silver phosphate, Ag_3PO_4, is a yellow precipitate produced when sodium phosphate and silver nitrate solution are mixed; and as some nitric acid is formed the liquid becomes acid:—

$$Na_2HPO_4 + 3AgNO_3 = Ag_3PO_4 + 2NaNO_3 + HNO_3.$$

Tricalcium phosphate, or bone ash, is $Ca_3P_2O_8$; it constitutes a large part of the earthy matter of bone, and is found in nature, in apatite, in coprolites, etc. It is useful as food for plants, but more so if treated with sulphuric acid to render the phosphoric acid soluble.

$$Ca_3P_2O_8 + 2H_2SO_4 = CaH_4P_2O_8 + 2CaSO_4.$$

By the addition of the right amount of sulphuric acid a monocalcium phosphate is produced, soluble in water, and therefore more readily available for use by plants. The product containing soluble phosphate and calcium sulphate is known as 'superphosphate,' and used as a fertiliser.

Phosphorous acid. H_3PO_3.

This acid is formed (1) by dissolving the anhydride (P_2O_3) in water:

$$P_2O_3 + 3H_2O = 2H_3PO_3.$$

(2) By decomposing trichloride of phosphorus by water:

$$PCl_3 + 3H_2O = H_3PO_3 + 3HCl.$$

(3) By allowing phosphorus slowly to oxidise in moist air.

Sticks of phosphorus are placed in test-tubes, each having a hole at the bottom; the tubes are supported in a funnel over a jar, and the whole standing in a dish of water covered with an open bell-jar. An acid liquid is gradually formed as the sticks of phosphorus oxidise in the moist atmosphere. The solution contains phosphorous and phosphoric acids. When liquid phosphorous acid is concentrated to a small bulk, it forms a syrup, which can be crystallised, although with some difficulty. By heating, the acid decomposes, thus:—

$$4H_3PO_3 = 3H_3PO_4 + PH_3$$

Phosphoric acid and phosphuretted hydrogen being formed. With alkalies and other metallic compounds **phosphites** are obtained. All phosphites decompose by heat, yielding phosphate, and sometimes phosphuretted hydrogen, or free hydrogen. From the fact that the phosphites absorb oxygen to change into phosphates, they act generally as reducing agents in metallic solutions and are able to throw down silver, mercury, copper, etc. in the metallic state.

Hypophosphorous acid. H_3PO_2.

No oxide corresponding to this acid is known, and the acid is obtained from a salt:—

$$CaH_4P_2O_4 + H_2SO_4 = CaSO_4 + 2H_3PO_2.$$

Thus, when calcium hypophosphite is treated with sulphuric acid, insoluble calcium sulphate is formed, and a solution of hypophosphorous acid. By cautious evaporation the acid may be obtained as a syrup and also in crystals, but it easily breaks up on heating thus:—

$$2H_3PO_2 = H_3PO_4 + PH_3.$$

Phosphoric acid and hydrogen phosphide are obtained.

Hypophosphites may be prepared by neutralising

Phosphine.

hypophosphorous acid with alkaline solutions; but are preferably made by boiling phosphorus itself in solutions of potash, barium, etc., as in the method for making hydrogen phosphide:—

$$3KHO + P_4 + 3H_2O = 3KH_2PO_2 + PH_3.$$

A solution of hypophosphite is obtained, from which the salt can be separated by evaporation. Hypophosphites, when heated, break up into phosphate and hydrogen phosphide, and from their capacity for taking up more oxygen, are used as reducing agents.

Phosphamine, H_3PO, or hydroxyl phosphine, is unknown, but an ethyl compound of analogous composition has been made $(C_2H_5)_3PO$.

Phosphine, or phosphuretted hydrogen, PH_3, is obtained by several reactions already mentioned. (1) from phosphorous acid and some phosphites; (2) from hypophosphorous acid and hypophosphites; (3) by action of phosphorus on alkaline solutions.

$$3KHO + P_4 + 3H_2O = 3KH_2PO_2 + PH_3.$$

Phosphine is a gas, and, when prepared by the third method, takes fire spontaneously in the air (fig. 47), owing to an admixture of traces of the liquid phosphide, P_2H_4. It has an unpleasant smell, 'ancient and fishlike,' and is slightly soluble in water.

Fig. 47.

Phosphine has slight basic characters, as it combines with gaseous hydrogen iodide, forming crystals of 'phosphonium' iodide, PH_4I, but the compound is decomposed by water into the constituents from which it is formed. It also unites with gaseous hydrogen chloride under pressure. Chlorine, bromine, and iodine all decompose phosphine by uniting with the hydrogen; and, like ammonia, it forms crystalline solid compounds with tin tetrachloride ($SnCl_4$) and some other metallic chlorides.

The alcoholic organic representatives of this body, tri-ethyl phosphine $P(C_2H_5)_3$ and $P(C_2H_5)_4OH$ tetrethyl phosphonium hydrate, corresponding to NH_3 and NH_4OH, are strongly alkaline; a solution of the latter phosphonium derivative, behaves almost like caustic potash.

CHAPTER XII.

Oxidized compounds of Chlorine, Bromine, Iodine.

THE halogen elements form various oxides which, in combination with water, yield acids, or in combination with bases yield salts. The following table shows the oxides and oxyacids derived from chlorine:—

Cl_2	chlorine.		
Cl_2O	chlorine monoxide.	$HClO$	hypochlorous acid.
Cl_2O_3	chlorine sesquioxide.	$HClO_2$	chlorous acid.
$[Cl_2O_4]\ ClO_2$	chlorine peroxide.		
$[Cl_2O_5]$?		$HClO_3$	chloric acid.
$[Cl_2O_7]$?		$HClO_4$	perchloric acid.

Oxides of Chlorine.

The oxides of chlorine are yellowish gases with a peculiar odour somewhat resembling that of the element: they are condensable by cold to the liquid state, but on heating they explode, and decompose into mixtures of free oxygen and chlorine, at the same time the yellow tint changes into the slightly green tint of diluted chlorine. In combination with water they produce acid liquids, which have powerful oxidizing and bleaching characters. They can in no instance be obtained by the direct union of oxygen gas with chlorine.

Chlorine protoxide, Cl_2O, is formed either by dehydration of hypochlorous acid, or by passing chlorine over mercuric oxide kept very cold:—

$$2HClO - H_2O = Cl_2O.$$
$$HgO + 2Cl_2 = HgCl_2 + Cl_2O.$$

With water this oxide forms **hypochlorous** acid, and with caustic alkaline solutions yields **hypochlorites**.

Chlorous anhydride. Chlorine sesquioxide. Cl_2O_3.

This very unstable body is obtained by reducing chloric acid :— $2HClO_3 + N_2O_3 = Cl_2O_3 + 2HNO_3$.

A mixture of potassium chlorate, nitric acid and arsenic trioxide is very gently heated: the arsenic oxide first reduces nitric acid, and the nitrogen trioxide thus formed further acts on the chlorate. The oxide is a yellow explosive gas, soluble in water forming **chlorous** acid; which combines with alkalies to form **chlorites**. [It has been asserted that this gas is a mixture, but the evidence is not conclusive.]

Chlorine peroxide. ClO_2. Vapour density, 33·65.

Peroxide of chlorine is liberated as a yellow gas when a chlorate is acted upon by an acid, e.g. sulphuric, hydrochloric or oxalic acid. It appears to result from the decomposition of chloric acid, thus:—

(i) $KClO_3 + H_2SO_4 = HClO_3 + KHSO_4$.
(ii) $3HClO_3 = HClO_4 + 2ClO_2 + H_2O$.

Chlorine peroxide is formed while part of the chloric acid is oxidized to perchloric acid. The following experiments will show the characters of the gas, but on account of its explosive character only small quantities of material should be used. (1) A few crystals of potassium chlorate in a test-tube are treated with strong sulphuric acid; the liquid turns yellow, and on the tube being brought into a flame a series of crackling explosions may be produced. Or if the tube (2) containing the mixture is placed in warm water, a yellowish gas will be given off and fill the tube; and on bringing a hot wire into the gas a loud detonation will follow. The yellow colour disappears, and the colour of the liberated chlorine mixed with oxygen is almost invisible. The explosion is somewhat violent, and sometimes, but not often, breaks the tube. With water or alkalies it breaks up into chlorite and chlorate.

Hypochlorous acid and hypochlorites.

Hypochlorites are produced by the action of chlorine or *cold dilute* solutions of alkalies, e. g.—

$$2KOH + Cl_2 = KCl + KClO + H_2O.$$

If chlorine is passed into solution of caustic potash, a chloride and a **hypochlorite** of the metal are formed. If dilute sulphuric or nitric acid be added to the liquid, in quantity sufficient only to decompose the hypochlorite and not the chloride, an aqueous solution of hypochlorous acid is obtained, and the acid can be distilled off and collected in a receiver in the liquid form.

Hypochlorous acid has a faint odour of chlorine oxide, and is an energetic oxidising agent. With hydrochloric acid it gives off chlorine—

$$HClO + HCl = H_2O + Cl_2.$$

It is an unstable compound, and readily decomposes, forming chloric acid, chlorine, and oxygen: solutions of hypochlorites, when simply boiled, are decomposed into chlorate and chloride.

$$3KClO = 2KCl + KClO_3.$$

Chloric acid and chlorates.

Alkaline chlorates are obtained by the decomposition of hypochlorites mentioned above, or by passing chlorine into a somewhat **strong** solution of alkali, and heating the liquid:—

$$6KOH + Cl_6 = 5KCl + KClO_3 + 3H_2O.$$

Chloride and chlorate are thus obtained, and the chlorate being much less soluble, is separated from the chloride by crystallisation.

Chloric acid may be obtained by acting on solution of barium chlorate with sulphuric acid, whereby insoluble barium sulphate is precipitated :—

$$BaCl_2O_6 + H_2SO_4 = 2HClO_3 + BaSO_4.$$

The dilute solution of the acid can be concentrated to a limited extent, but beyond a certain point it decomposes into perchloric acid, etc.

Potassium chlorate is generally used for the preparation of oxygen gas, and for many operations is a useful oxidising substance. When fused, it yields oxygen, and is employed in the preparation of manganates, chromates, etc. Mixed with organic substances, it produces very combustible or even explosive mixtures. It is used in the composition upon the tops of safety matches, and when rubbed, even slightly, upon phosphorus, ignition takes place.

If a small quantity of potassium chlorate be mixed with powdered sugar, and a single drop of sulphuric acid is brought with a glass rod upon the mass, the mixture instantly takes fire.

If a fragment of phosphorus, with a few crystals of chlorate, be dropped into a beaker of water, and a little strong sulphuric acid slowly poured through a thistle funnel upon the mixture: the phosphorus will oxidise and burn under water with flashes of light.

All chlorates are soluble in water, and accordingly no precipitates are obtained with reagents in testing for them. They may be recognised, however, by the evolution of oxygen when heated, and by the fact that the residue, after heating, when tested with silver nitrate, gives a precipitate of silver chloride. Their behaviour also with sulphuric acid is very characteristic.

Perchloric acid and perchlorates :—

These are the most stable oxidised compounds of chlorine, and the lower members of the series—hypochlorites, chlorites, and chlorates—have a tendency to pass into perchlorates, under suitable conditions. Let a few grammes of potassium chlorate be heated in a hard glass test-tube; at first the salt fuses (at 334°), and (at 354°) gives off oxygen freely, the melted salt appearing as if boiling. After a time the salt begins to solidify, and can hardly be kept liquid. This is due to the fact that perchlorate is formed along with chloride, and both salts fuse at a much higher temperature: the perchlorate also requires a greater heat to separate its oxygen.

$$2KClO_3 = KClO_4 + KCl + O_2.$$

The chloride and perchlorate can be separated by recrystallisation (potassium chloride being more soluble), and any admixture of chlorate decomposed by digesting with hydrochloric acid, which does not act on the perchlorate. It has already been mentioned that perchlorate is obtained by the action of strong acids on chlorate (see ClO_2).

To separate **perchloric acid**, the crystallised potassium perchlorate is mixed in a retort with excess of strong sulphuric acid, and distilled; a fuming colourless acid liquid condenses in the receiver, which is perchloric acid, $HClO_4$. It is a powerful oxidising agent, and sets fire to wood, paper, and other organic substances; when dropped upon charcoal it decomposes with a loud explosion. With water it forms crystals, H_3ClO_5.

It is found that pure potassium perchlorate, when heated, becomes partially reconverted into chlorate.

Bromine oxides and acids—

No oxide of bromine has yet been obtained, and **hypobromous** acid, HBrO, and **Bromic** acid, $HBrO_3$, are the only acids at present known: they are analogous to hypochlorous and chloric acids. Similarly, the **hypobromites** and **bromates** resemble the corresponding chlorine compounds in composition and general chemical properties.

Iodine oxides and acids—

The affinity of iodine for oxygen is very great, and consequently these two elements form compounds of a very stable character, and differing in a marked way from the explosive unstable oxides of chlorine. And in respect of its oxygen compounds, iodine can expel chlorine, although the latter element has a far greater affinity for hydrogen and metals (page 92).

Iodic acid, HIO_3, is formed by boiling iodine with nitric acid, and thus directly oxidising it; or by passing excess of chlorine into water containing solid iodine in suspension :—

$$I + Cl_5 + 3H_2O = HIO_3 + 5HCl.$$
$$(* cp.\ PCl_5 + 3H_2O = HPO_3 + 5HCl.)$$

Crystals of iodic acid are obtained by evaporating the solution (after removal of hydrogen chloride); and by exposure of these crystals to a temperature of 200° water is given off and **Iodine pentoxide**, I_2O_5, is left. This oxide, when strongly heated, is decomposed into its component elements.

Iodine trioxide, I_2O_3, has been produced by oxidising the vapour of iodine by means of ozone.

Potassium iodate may be obtained (i) by the action of iodine upon potash solution :—

$$6KOH + 6I = KIO_3 + 5KI + 3H_2O.$$

This method resembles the formation of the chlorate, except that no lower compound corresponding to an hypochlorite can be obtained.

(ii) By the action of iodine with chlorine upon potash solution:—

$$6KOH + (I + Cl_5) = KIO_3 + 5KCl + 3H_2O.$$

The iodine is dissolved in the potash, and chlorine gas passed in; no chlorate is obtained, as the affinity of iodine for oxygen is stronger than that of chlorine.

(iii) By decomposing potassium chlorate with iodine.

$$2KClO_3 + I_2 = 2KIO_3 + Cl_2.$$

(iv) By neutralising iodic acid with potassium hydrate or carbonate.

Iodates break up if strongly heated, giving off oxygen, and in some instances iodine. In aqueous solutions they are reduced by deoxidising agents, such as sulphur dioxide, hydrogen sulphide, etc. Hydriodic acid reduces iodic acid with a separation of iodine.

$$HIO_3 + 5HI = 3I_2 + 3H_2O.$$

Periodic acid may be obtained by acting on iodine with perchloric acid, and the crystals have the composition H_5IO_6 ($I_2O_7 + 5H_2O$). The acid HIO_4 is not known. When heated to 160° the acid decomposes, giving off water and oxygen, leaving iodic anhydride.

$$2H_5IO_6 = 5H_2O + I_2O_5 + O_2.$$

The existence of periodic anhydride, I_2O_7, is doubtful.

Potassium periodate is formed by oxidising iodate with chlorine in presence of potash, probably by the action of hypochlorite which would first be formed.

$$KIO_3 + 2KOH + Cl_2 = KIO_4 + 2KCl + H_2O.$$

The periodates are remarkable for the variation in the amount of basic oxide associated with the acid oxide in the salt: the following will illustrate this variation:—

KIO$_4$ potassium periodate.
Ag$_3$IO$_5$ silver periodate.
Ba$_5$(IO$_6$)$_2$ barium periodate.

They may be considered as produced from a typical periodic acid, H$_7$IO$_7$, by the removal of water in successive stages.

Heats of formation.

The oxidised compounds of chlorine and iodine differ remarkably in respect of their stability, and iodine is able to expel chlorine from chlorates and perchlorates, although chlorine displaces iodine from all iodides, etc. The measurement of the heats of formation of these compounds gives a quantitative expression for these differences of chemical attraction.

The following table shows the heat of formation of the halogen acids and oxyacids in solution in water :—

R	(H, R, Aq)	(H, R, O$_3$ Aq).
Cl	39320 heat units.	23940 heat units.
Br	28380 ,,	12420 ,,
I	13170 ,,	55710 ,,

We see from these values that the conversion of HCl.Aq into HClAq.O$_3$ *absorbs* 15380 heat units, and HBr about the same quantity, but the oxidation of HI.aq to iodic acid produces a *development* of 42540 units of heat.

Fluorine. F. Atomic weight, 19.

Fluorine in many respects resembles in chemical characters the halogen elements, and is accordingly grouped with them; but until the end of 1886 the element had never been isolated, and its properties are still little known. This alone of all elements does not form any compound with oxygen.

Fluor spar, or calcium fluoride, is found as a crystallised mineral in Derbyshire and other localities. Cryolite, a double fluoride of aluminium and sodium, Na_3AlF_6, is found massive in Greenland. Fluorine is found as a constituent in some other minerals, and in small quantities appears widely distributed in nature.

Hydrogen fluoride, or hydrofluoric acid, HF, is obtained by the action of strong sulphuric acid on fluor spar. As the acid attacks glass, the operation is performed in vessels made either of lead or platinum.

$$CaF_2 + H_2SO_4 = CaSO_4 + 2HF.$$

The gas evolved is very soluble in water, in which it can be collected, or it may be liquefied in a freezing mixture; as the acid is very corrosive it must be most carefully dealt with. Dilute hydrofluoric acid can be kept in gutta-percha bottles, but the liquid anhydrous acid can only be kept in platinum vessels.

By the electrolysis of the anhydrous hydrogen fluoride (obtained by heating potassium and hydrogen fluoride KHF_2), Monsieur Moissan has succeeded in isolating the element fluorine. The acid having been cooled to $-23°$ in a platinum U-tube, and a current of twenty Bunsen cells passed through the liquid, hydrogen is liberated at one pole and fluorine gas from the other. The gas manifests the most remarkable chemical activity, and the elements silicon, boron, arsenic, antimony, sulphur, and iodine take fire in it instantly. Organic bodies, such as alcohol, ether, turpentine, also inflame directly they are brought in contact with the gas.

Water is decomposed, hydrogen fluoride, oxygen, and ozone being formed. If mixed with hydrogen, even in the dark, the gases combine with explosion.

The vapour density of hydrogen fluoride has been found to be ten times that of hydrogen: its molecular weight therefore is 20. From the amount of silver obtained from silver fluoride when that salt is decomposed in hydrogen, the **atomic weight of fluorine** has been found to be 19.

The most characteristic property of fluorine is its great chemical affinity for silicon, and hydrofluoric acid quickly attacks silicates, forming gaseous silicon fluoride:—

$$SiO_2 + 4HF = SiF_4 + 2H_2O.$$

Solution of hydrofluoric acid or the vapours are much used for etching glass. The experiment can readily be shown thus:—A watch-glass is prepared by coating the convex side with bees-wax, and some portion of the glass laid bare by writing with a pointed instrument so as to scratch through the wax. A little fluor spar is placed in a leaden (or platinum) cup, moistened with strong sulphuric acid, and covered with the watch-glass; a very gentle heat may be applied, but the wax must not be melted. After a time, if the wax is cleaned off, the writing on the glass will be found to be etched into the surface.

The following list shows the composition of some important fluorides:—

HF,	hydrogen fluoride.
KHF_2,	potassium and hydrogen fluoride.
CaF_2,	calcium fluoride.
BF_3,	boron fluoride.
KBF_4,	potassium boro-fluoride.
SiF_4,	silicon fluoride.
K_2SiF_6,	potassium silico-fluoride.
PF_5,	phosphorous penta-fluoride.
CrF_6,	chromium hexa-fluoride.
MnF_7,	manganese hepta-fluoride.

CHAPTER XIII.

BORON AND SILICON.

Boron. B. Atomic weight, 11.

Boron is an element of limited distribution in nature; it is found as boracic acid or borax in several parts of the world. A crude form of borax (known as Tincal) is found in Thibet, and a borax lake in California yields a large supply.

To obtain the element either the oxide B_2O_3 or the fluoride BF_3 may be decomposed with metals. When boron trioxide is strongly heated with metallic potassium or sodium an **amorphous** form of boron is produced; the double fluoride of potassium and boron, KBF_4, is also similarly reduced by potassium.

Amorphous boron thus obtained is a brown powder; it will combine directly with oxygen, sulphur, nitrogen, chlorine, and we may now add fluorine. When boron oxide is heated with aluminium, an allotropic variety is obtained in extremely hard crystals, and known as **adamantine** or diamond boron. Amorphous boron is also changed into this variety when heated with aluminium.

The crystals contain traces of aluminium and sometimes carbon; they are infusible and almost incombustible in oxygen, but will burn in chlorine and fluorine. The most important compounds of boron with non-metallic elements are :—

Boron hydride	BH_3.	Boron nitride	BN.
Boron fluoride	BF_3.	Boron trioxide	B_2O_3.
Fluoboric acid	HBF_4.	Boracic acid	$B(OH)_3$.
Boron chloride	BCl_3.		

Boron chloride, BCl_3, is formed when boron is heated in chlorine, and also when the oxide is acted upon at a red heat by chlorine and carbon simultaneously.

$$B_2O_3 + 3C + 3Cl_2 = 2BCl_3 + 3CO.$$

It is a fuming liquid resembling phosphorous trichloride: it boils at 17°, and distils without decomposition. Its vapour density is 58.7, agreeing with the formula BCl_3 [molecular weight 117.5].

Water decomposes it into boracic acid and hydrochloric acid.

$$BCl_3 + 3H_2O = B(OH)_3 + 3HCl.$$

Boracic acid is present in the steam issuing from the earth in volcanic districts of Tuscany: by condensing the steam in open basins the boracic acid is also condensed and may be separated by evaporation of the liquid.

In the laboratory the acid is made from borax. A warm saturated solution of borax is made acid with excess of hydrochloric acid, and as the liquid cools boracic acid separates in crystalline scaly plates. It is readily soluble in hot water, but only sparingly in cold. When dried the solid acid has a greasy feel between the fingers like soap-stone.

Boracic acid (or boric acid) is a feebly acid substance, it is almost tasteless, and its solution in water, like carbonic acid, changes litmus to a wine-red colour. Yellow turmeric paper is turned brown by the acid, a characteristic which is employed in testing for boracic acid.

Boracic acid, and its alcoholic solution (which contains ethyl borate), and also boron fluoride when brought into the flame of a Bunsen burner colour the flame bright green. The acid is an **antiseptic**, and is sometimes used as a preservative for milk, fish, and perishable articles of food.

When heated to 120° boracic acid loses water, and is converted into metaboric acid HBO_2 :—

$$B(OH)_3 - H_2O = BO(OH)$$

which at a higher temperature loses more water and boric anhydride—boron trioxide is left.

$$2HBO_2 - H_2O = B_2O_3.$$

Pyroboric acid, $H_2B_4O_7$, may be obtained as an intermediate product by heating the acid less strongly.

Boron trioxide, B_2O_3, is a white substance, fusible below redness, and solidifying into a glassy mass. It slowly dissolves in water forming boracic acid.

Borates. In combination with alkaline and other bases boracic acid forms the salts known as borates. They are somewhat complex in character, but may be arranged under three typical forms :—

Orthoborates	M_3BO_3	$B_2O_3 + 3H_2O.$
Metaborates	MBO_2	$B_2O_3 + H_2O.$
Pyroborates	$M_2B_4O_7$	$2B_2O_3 + H_2O.$

Common **borax** or sodium biborate is referred to the last type. The crystallised salt has the composition $Na_2B_4O_7, 10H_2O$, but when heated the water is expelled and a glassy mass $Na_2B_4O_7$ is left. Borax swells in a remarkable manner when heated, but after all the water is expelled, passes into a state of quiet fusion. The fused borax glass has the property of dissolving metallic oxides, and forms with them coloured glasses, and the characteristic colours thus produced are useful as blowpipe tests in analysis. Thus a bead of fused borax is coloured blue by cobalt, green by chromium, yellow by iron peroxide, etc.

A hydride of boron (BH_3 ?) has been obtained, but

not in a state of purity; an organic compound with ethyl—Boron triethide—$B(C_2H_5)_3$ is known; it is analogous in composition to triethylamine $N(C_2H_5)_3$ and triethylphosphine $P(C_2H_5)_3$.

Boron fluoride. BF_3. Vapour density, 34. This compound is obtained as a gas by acting on boron trioxide with hydrogen fluoride.

$$B_2O_3 + 6HF = 2BF_3 + 3H_2O.$$

It is very soluble in water: on dilution the solution decomposes, forming boric acid and fluoboric acid.

$$4BF_3 + 3H_2O = 3HBF_4 + B(OH)_3.$$

Silicon. Si. Atomic weight, 28.

Although silicon is, next to oxygen, the most abundant element, and forms so large a portion of the earth's crust, the properties of the element by itself are comparatively little studied, and it has never been prepared on a large scale.

Silicon oxide, SiO_2, or silica, is found as quartz, rock-crystal, flint, chalcedony, agate, jasper, cairngorm and sand; and, in combination with bases, it is a constituent of an infinite variety of minerals. Most natural waters contain dissolved silica.

The element silicon is separated from the fluoride by metals, and is known in two allotropic states. Potassium silicofluoride is heated with metallic potassium, and after washing a brown **amorphous** powder is left.

$$K_2SiF_6 + 4K = Si + 6KF.$$

But if the silicofluoride is decomposed with an alloy of zinc and potassium, the silicon is obtained in **crystals** hard enough to scratch glass. The density of these crystals is 2·49, which is less than that of diamond.

Silica.

In the amorphous state silicon is a combustible powder, soluble in potash or aqua regia, but if ignited strongly in a closed vessel it becomes more dense and much more difficult to act upon by chemical agents. It will not burn nor dissolve in any acids, except a mixture of nitric and hydrofluoric acids.

Silicon is a **tetrad** element, since it combines with four atoms of hydrogen, chlorine or fluorine, and its compounds resemble in composition those of carbon and titanium. The following are some of the chief typical compounds:—

Silicon hydride	SH_4.	Silicon fluoride	SiF_4.
Silicon tetra-ethide	$Si(C_2H_5)_4$.	Hydro fluo-silicic acid	H_2SiF_6.
Silicon tetra-chloride	$SiCl_4$.	Silicon oxide	SiO_2.
Silicon trichloride	Si_2Cl_6.	Meta silicic acid	H_2SiO_3.
Silicon chloroform	$SiHCl_3$.	Ortho silicic acid	H_4SiO_4.

Ethyl ortho-silicate $(C_2H_5)_4SiO_4$.

Silica or Silicon dioxide. SiO_2.

This is the only oxide of silicon, and is similar in composition to CO_2; no lower oxide exists, but the hydrate silicoformic acid is known, H_2SiO_2 (H_2O, SiO), analogous to formic acid H_2CO_2, (or H_2O, CO).

Crystalline silica is found as quartz, rock-crystal, and sand in variously coloured varieties: the amorphous forms are flint, chalcedony, opal and others; in the last-named a recognisable amount of water is found. The density of quartz is about 2·6; it is very hard, but not so hard as ruby, corundum (Al_2O_3) and diamond. All forms of silica are more or less soluble in alkali by prolonged heating, but crystalline quartz is the least attacked.

Silica is artificially obtained from silicic acid, which by heating loses all combined water. From gelatinous silicic acid it is left as light white powder, soluble in alkalies.

Silicic acids.

When silica mixed with alkaline carbonate is brought into a state of fusion at a high temperature, carbonic acid is expelled and a silicate formed, carbon dioxide escaping with effervescence from the fused alkali.

$$Na_2CO_3 + SiO_2 = Na_2SiO_3 + CO_2.$$

If the silica be used in excess, an acid silicate or glass is formed which remains transparent on cooling; but basic silicates, containing excess of alkali, are opaque when cold.

Alkaline silicates produced by fusing sand with *excess* of alkali are soluble in water, forming alkaline solutions; and from such solutions, by the addition of hydrochloric acid, silicic acid is separated, and after a short time becomes a jelly-like mass.

A solution of silicic acid in water can be obtained by *dialysis* in the following manner. An aqueous solution of silicate of soda is poured into excess of hydrochloric acid, sodium chloride is formed, which with silicic acid remains in solution. The liquid is *dialysed* by being placed in a vessel, with a membranous bottom of vegetable parchment, and the whole floated in a quantity of pure water (fig. 48).

Fig. 48.

The parchment permits the passage only of crystalline bodies, and so the salt and acid pass out into the water; but gelatinous substances (or *colloids*) cannot pass through such a membrane, and the silica is retained; by repeatedly changing the water an aqueous solution of silicic acid is left pure. The solution may be concentrated by boiling in a flask to a certain extent (14 per cent. is about the limit), but after a time it gelatinises;

probably the solution contains orthosilicic acid (H_4SiO_4), but evaporated in vacuo the residue has the composition H_2SiO_3.

Silica is perceptibly soluble in water, and is present in most natural waters. Many hot volcanic springs contain a considerable amount in solution, which deposits as a 'sinter' when the water cools. Of this nature are the rock-basins of geysers, and the beautiful pink and white terraces in New Zealand lately destroyed by volcanic eruptions. It has been shown that the presence of silica in the water of a town supply, in a quantity of about one grain per gallon, has the effect of preventing lead pipes from being attacked. Certain soft waters act upon lead and dissolve quantities sufficiently large to be dangerous to health, while other soft waters do not; and the difference appears due to the presence of a larger quantity of dissolved silica in the latter, which, by forming an insoluble surface of lead-silicate, coats the pipes and stops the harmful solvent action of the water.

Solutions of silicate, when added to metallic salts, e.g. copper, iron or alumina, give insoluble precipitates of metallic silicates.

The composition of natural silicates is complex, but for the most part they fall under the following typical forms of silicic acid:—

H_4SiO_4 orthosilicic acid or $2H_2O$ SiO_2.
H_2SiO_3 metasilicic acid or H_2O SiO_2.
$H_6Si_2O_7$ parasilicic acid or $3H_2O$ $2SiO_2$.
$H_4Si_3O_8$ sesquisilicic acid or $2H_2O$ $3SiO_2$.

The following minerals may be taken as examples:—

Mg_2SiO_4, olivine; $MgSiO_3$, augite; $Al_2Si_2O_7$, clay; $K_2Al_2Si_6O_{16}$, felspar.

Glass is a material of which the composition varies according to the purpose for which it is required. The principal kinds are crown, or window-glass; flint, or white bottle-glass, green bottle-glass; **hard** glass for combustion tubing.

For all varieties silica is of course the acid constituent; the bases used are, (1) for window-glass, lime and soda; (2) for flint-glass, in tumblers, bottles, etc., **lead** and potash; (3) for green bottle-glass, soda with lime, alumina and iron; (4) for combustion tubing, potash and lime. The proportions vary considerably for all kinds in different glass-works.

Silicon Chloride, $SiCl_4$, is produced, like boron chloride, by the joint action of chlorine and carbon upon the oxide.

A mixture of silica, with oil and carbon is made into balls and heated to bright red heat while a current of chlorine is passed through the apparatus:—

$$SiO_2 + C_2 + Cl_4 = SiCl_4 + 2CO.$$

The silicon chloride is condensed in a tube kept cold by a freezing mixture. It is a fuming liquid, instantly decomposed by water into silicic and hydrochloric acids.

$$SiCl_4 + 4H_2O = Si(OH)_4 + 4HCl.$$

It boils at 50°, and the density of the vapour is 85.

Silicon fluoride, SiF_4. Vapour density 52. This compound is formed whenever hydrofluoric acid acts upon silica or a silicate, the attraction of fluorine for silicon being remarkably great.

To prepare the gas in the laboratory a mixture of sand and fluor spar is placed in a thick glass bottle, such as a soda-water bottle, and strong sulphuric acid added. The bottle is then placed in a saucepan of cold water or upon sand, which is gradually heated until a steady stream of gas is given off.

$$2CaF_2 + SiO_2 + 2H_2SO_4 = 2CaSO_4 + SiF_4 + 2H_2O.$$

Silicon Fluoride.

The gas fumes strongly in moist air and unites with gaseous ammonia to form a white solid.

Silicon fluoride is decomposed in a remarkable manner by water, to show which the gas should be made to bubble through a little mercury in a glass so as to pass upwards into the water covering it (fig. 49). Each bubble of gas, as it comes in contact with the water, decomposes,

Fig. 49.

forming a skin of gelatinous silica. But at the same time hydrofluosilicic acid forms, which remains dissolved.

$$4H_2O + 3SiF_4 = Si(OH)_4 + 2H_2SiF_6.$$

Orthosilicic acid and hydrofluoric acid will first be formed, and the latter, by its union with more silicon fluoride, forms hydrofluosilicic acid—

$$(2HF, SiF_4 = H_2SiF_6).$$

Silicon hydride, SiH_4, is given off as a spontaneously inflammable gas on dissolving magnesium silicide in acid:—

$$Mg_2Si + 4HCl = 2MgCl_2 + SiH_4.$$

The gas is not pure, but contains some hydrogen.

CHAPTER XIV.

Arsenic. As. Atomic weight, 75.
Vapour density, 150. Molecular weight, $300 = As_4$.

ARSENIC is associated closely with phosphorus on one side and with antimony on the other by the similarity of the compounds with hydrogen, chlorine and sulphur, and the characters of the oxides and acids.

Arsenic is found in the uncombined state in mineral veins, and also associated with metals as arsenides and sulpharsenides. It is commonly present as an impurity in mineral sulphides and ores. Realgar is a sulphide, As_2S_2; orpiment is As_2S_3; mispickel or arsenical iron is Fe As S; nickel arsenide, Ni As. A few mineral arsenates are found.

The element arsenic is obtained in the free state by heating mispickel in closed vessels—

$$4\,FeSAs = 4\,FeS + As_4.$$

Or it may be obtained by the reduction of the oxide As_2O_3 with carbon.

Arsenic is a brittle substance, somewhat crystalline, with a little lustre; it volatilizes without fusion at low temperatures. The density of the solid is 5·9, and the vapour-density of the element is 150. Wherefore (as in the case of phosphorus) the vapour molecule is represented by As_4. On account of the poisonous character of arsenical substances great care and caution are necessary in experiments with them.

The following are some of the most important and

typical arsenic compounds; and their general similarity to the corresponding phosphorus compounds will be noticed:—

Arsenic trioxide	As_2O_3.	Arsenic trisulphide	As_2S_3.
Arsenic pentoxide	As_2O_5.	Arsenic pentasulphide	As_2S_5.
Arsenic hydride	AsH_3.	Arsenic chloride	$AsCl_3$.
Silver arsenite	Ag_3AsO_3.	Potassium sulpharsenite	K_3AsS_3.
Silver arsenate	Ag_3AsO_4.	Potassium sulpharsenate	K_3AsS_4.

Arsenic Trioxide, As_2O_3, is the substance commonly known as white arsenic. It is obtained by oxidation or roasting of arsenical ores, e.g. mispickel FeAsS. The crude product is sublimed to purify it, and obtained as a glassy, almost transparent mass known as **vitreous** arsenic; this after a while changes to an **opaque allotropic** form of slightly different density. The oxide is obtained in **crystals** by sublimation:—a small fragment of white arsenic is placed in a test-tube and gently heated in a flame; the trioxide sublimes without melting, and condenses in brilliant crystals (octahedra) (fig. 50) in a cooler part of the tube.

Fig. 50.

Arsenic trioxide is with difficulty soluble in water, more soluble in hydrochloric acid, and readily soluble in solution of alkalies or alkaline carbonates, forming **arsenites**. The strong solution of arsenic trioxide in hydrochloric acid will deposit brilliant transparent octahedral crystals.

The vapour-density of arsenic trioxide is found to be 198: the molecular formula is therefore As_4O_6. To show the easy reduction of the oxide, heat a little of the powder with a fragment of dry charcoal in a tube,

L

when a black mirror of sublimed arsenic will be obtained.

Arsenious acid is not known in a definite condition, although doubtless formed in the solution of the oxide, but metallic arsenites are numerous:—

$$KH_2AsO_3, \quad CuHAsO_3, \quad Ag_3AsO_3$$

may be taken as examples.

Arsenic pentoxide As_2O_5 is obtained from the lower oxide by heating it with nitric acid, and evaporating off excess of water and acid: a syrupy liquid containing **arsenic acid** (H_3AsO_4) is left, which by further heating is resolved into water and arsenic pentoxide:—

$$2H_3AsO_4 = As_2O_5 + 3H_2O.$$

The pentoxide is a white amorphous solid, dissolving in water, to reproduce arsenic acid: at a strong red heat it loses oxygen and the trioxide is formed.

Arsenic acid, when neutralised with alkalies, forms **arsenates** similar in character to orthophosphates; they have the same crystalline form and general properties, and are obtained by similar reactions.

Arsenate of silver Ag_3AsO_4 is produced as a brick-red precipitate by mixing sodium arsenate with silver nitrate:—

$$3AgNO_3 + Na_2HAsO_4 = Ag_3AsO_4 + HNO_3 + 2NaNO_3.$$

With magnesia mixture, arsenates precipitate the double arsenate of magnesium and ammonium—$MgNH_4AsO_4, 6H_2O$ resembling the phosphate.

Arsenic sulphides. As_2S_3 As_2S_5 are made by the action of hydrogen sulphide upon acid solutions of arsenites and arsenates, being thrown down as yellow precipitates. They dissolve in alkaline sulphides, forming **sulpharsenites** and **sulpharsenates**, some of which are crystallizable salts.

Arsenic Hydride.

Arsenic hydride or Arsine, AsH_3. This compound, analogous to ammonia and phosphine, has not been obtained pure, but is given off admixed with hydrogen on dissolving zinc arsenide in hydrochloric acid.

$$Zn_3As_2 + 6HCl = 3ZnCl_2 + 2AsH_3.$$

It is a dangerously poisonous gas with an odour of garlic; it is combustible, burning with a dull bluish-coloured flame, forming arsenious oxide and water. It appears to have no basic characters; no combinations with acids have been obtained, and the solution in water is not even alkaline to test-paper. The organic representatives—trimethyl arsine $(CH_3)_3As$, etc.—are alkaline, and readily combine with acids.

Marsh's test for Arsenic. This test depends on the fact that arsenic hydride is formed when nascent hydrogen acts on a compound of arsenic:—

$$H_3AsO_3 + 3H_2 = AsH_3 + 3H_2O.$$

A small flask filled with thistle funnel and a delivery tube, as for the production of hydrogen, is used (fig. 51): pure zinc and hydrochloric acid are introduced, and after a short time the hydrogen is ignited. It is advisable to cover the flask with a cloth before igniting the gas, as an explosion may happen unless the air has all been driven out. If the materials are pure the hydrogen flame gives no deposit upon a piece of cold porcelain brought into it, but commercial zinc usually contains arsenic. When the purity of the gas is proved, a little solution of arsenite may be poured down the thistle funnel, which will produce a more rapid evolution of gas, and the flame will become larger and perceptibly coloured. A piece of cold porcelain depressed upon the flame will be covered with a deposit of metallic arsenic. The films of arsenic are metallic looking in the thicker places, brownish near the edges: they are easily volatilized by heat, and dissolve in solution of bleaching powder. (Compare the behaviour of antimony.) A portion of the glass tube from which the gas is burnt should be heated to redness; the gas decomposes, and a deposit of arsenic appears on the tube, which may be identified in a similar way, or may be converted into

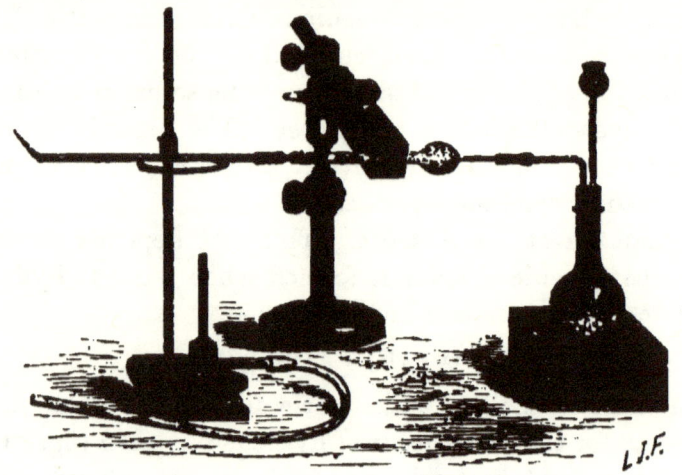

Fig. 51.

crystals of oxide by cautious sublimation in an open tube.

Reinsch's test for Arsenic is made by heating a piece of bright copper sheet in a hydrochloric acid solution of arsenical substance. The deposition of a black crust on the copper results from the precipitation of arsenic in the free state. The copper, with its deposit, is dried at a gentle heat, and placed in a small glass tube, which is then gently heated in a flame. Sublimation and union with oxygen take place, and a ring of brilliant crystals of trioxide forms upon the side of the tube.

The reduction of arsenic from its oxide can also be

Arsenic. 149

shown by making a mixture of arsenic trioxide with dry potassium cyanide, and heating it in a glass tube.

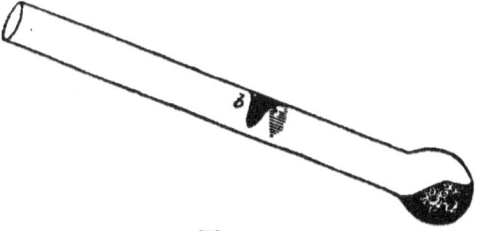

Fig. 52.

Arsenic is liberated and deposits as a bright metallic mirror (fig. 52) in the cooler part of the tube.

CHAPTER XV.

QUANTIVALENCE. ACIDS, BASES, AND SALTS.

In the compounds already studied there are many differences to be noticed in the *numbers* of atoms with which elements combine. Thus in hydrogen chloride HCl *one* atom of hydrogen unites with one of chlorine, but in water H_2O we find *two* atoms of hydrogen associated with an atom of oxygen: in ammonia NH_3 we find that a single atom of nitrogen can unite with *three* of hydrogen, and in marsh gas CH_4 the one carbon atom unites with *four* hydrogen atoms. Several terms are in use to express these differences, which are described as differences in **quantivalence, valency,** or **atomicity** of the elements. And the elements themselves are called **monad, dyad, triad,** or **tetrad,** according to the number of hydrogen (or other atoms) with which they associate.

This is illustrated by the following examples:—

Hydrogen is the type of a **monad** element, and the halogen elements in combination with it are all monad. Thus we have the compounds HF, HCl, HBr, HI formed by union of one atom of each. The compounds of either hydrogen or chlorine, or in some cases oxygen also, are taken to show the atomicity or quantivalence of other elements. Thus oxygen and sulphur are regarded as **dyads** in such combinations as H_2O, H_2S, Cl_2O. Boron, nitrogen, phosphorus and arsenic are **triads** in the following instances, BCl_3, NH_3, PH_3, $AsCl_3$. Carbon and silicon are manifestly **tetrads** in CH_4, CCl_4, $SiCl_4$, as well as in CO_2 and SiO_2. Again nitrogen, phosphorus and arsenic appear as **pentads** in such compounds as NH_4Cl, PH_4I, PCl_5, As_2O_5.

Sulphur is a **hexad** in SO_3 and $SO_2(OH)_2$, although **tetrad** in SO_2 and **dyad** in SH_2.

There are some elements which appear to have heptad characters, for example, chlorine and iodine in perchlorates and periodates. The oxides corresponding to these salts are Cl_2O_7 and I_2O_7, but they are not known in the free state: the free acids are $HClO_4$ or $ClO_3,(OH)$ and H_5IO_6 or $IO(OH)_5$, but both are derived from the typical acids $Cl(OH)_7$ and $I(OH)_7$ which are heptad, and by loss of water produce the ordinary forms.

That the quantivalence of elements is not a fixed and invariable character is shown by the examples of sulphur, chlorine, nitrogen, etc. The metal molybdenum furnishes a striking example of varying atomicity in the chlorides $MoCl_2$, $MoCl_3$, $MoCl_4$, $MoCl_5$. Nitrogen in the oxide NO, and carbon in CO furnish examples of doubtful cases, in which it seems impossible to settle the atomicity of either element.

When the elements are arranged in a series, as in the following table, in the order of their atomic weights, there is a manifest repetition or recurrence of type after the element fluorine. The first horizontal line contains seven elements, beginning with a monad metal lithum and ending with fluorine (as yet of doubtful atomicity), and the next gives us a second series almost exactly parallel in chemical character with the first. This periodic repetition of properties was discovered by Mr. J. A. R. Newlands.

Table of Elements to show the Periodic Law.

I.	II.	III.	IV.	V.	VI.	VII.	VIII.
R_2O	R_2O_2	R_2O_3	R_2O_4	R_2O_5	R_2O_6	R_2O_7	
H 1							
Li 7	Be 9	B 11	C 12	N 14	O 16	F 19	
Na 23	Mg 24	Al 27	Si 28	P 31	S 32	Cl 35·5	
K 39	Ca 40	Sc 44	Ti 48	V 51	Cr 52	Mn 55	Fe 56, Ni 58·6, Co 59
Cu 63	Zn 65	Ga 69	Ge 70	As 75	Se 79	Br 80	
Rb 85	Sr 87	Yt 89	Zr 90	Nb 94	Mo 96	—	Rh 104, Ru 104·5, Pd 106
Ag 108	Cd 112	In 114	Sn 118	Sb 120	Te 126	I 127	
Cs 133	Ba 137	La 139	Ce 141	Di 144	Sm 150	—	
—	—	—	Ho 160	Er 166	Yb 173	—	
—	—	—	—	Ta 182	W 184	—	Ir 192·5, Os 193, Pt 194·5
Au 196	Hg 200	Tl 204	Pb 207, Th 232	Bi 209, U 240			

The parallel is not so exact as the atomic weights increase, but the general characters of the first series of elements are repeated in the others. In column I we have the monad metals, in II the dyads, and so on, all natural groups being brought together. This repetition of type is known as the **periodic law** of the elements. In the last column three groups of metals appear which do not fit otherwise into the arrangement; they are the iron group, palladium group, and the platinum group. The elements gallium, Ga 69, and germanium, Ge 70, were unknown when the table was first drawn up, but their probable existence was predicted by the distinguished Russian chemist Mendelejeff, and he also foretold their general characters and atomic weights with what afterwards proved a near approach to the reality. And it may be that other blanks at present existing in the table will be filled hereafter by the discovery of new elements.

Acids. We have seen that certain non-metallic elements form oxides, which, by addition of water, produce sour or acid substances, and these acid bodies will react with metallic oxides or hydrates to form neutral compounds or salts. Such are the **oxygen acids**, or oxyacids, of which sulphuric and phosphoric acid are good examples. But the halogen elements form a group of acids of a different type; they unite first with hydrogen, and these compounds combining with water form powerful acids, with sour taste, and the power of forming salts. Hydrochloric acid, hydrofluoric acid, etc., belong to this class; and we may add fluoboric acid HBF_4, and fluosilicic H_2SiF_6.

The usual characters by which an acid is recognised are sourness and the power of changing the colour of blue litmus to red. But these characters are manifested to a slight extent only by some compounds, such as carbonic acid, boracic acid and silicic acids. Yet chemically we regard these as true acids on account of their power of producing salts if combined with bases.

When an acid is changed into a salt it loses hydrogen which becomes replaced by a metal, and the hydrogen in an acid may be **partly** or **wholly** removed in the formation of salts.

The term **basicity** is used to express differences in this respect shown by various acids.

Thus hydrogen chloride has *one* atom of hydrogen only in the molecule, and forms with **bases** (metallic oxides) but one chloride; it is therefore termed *mono-*

basic. Hydrobromic and hydriodic acids are monobasic like hydrochloric acid.

Another monobasic acid is nitric acid, HNO_3, which neutralised with soda becomes sodium nitrate, $NaNO_3$. We sometimes express this monobasic character of the acid by writing the formula $NO_2(OH)$, and the salt similarly is $NO_2(ONa)$.

A **bibasic** acid contains **two** atoms of hydrogen replaceable by metals, and is able to form both neutral and acid salts. Carbonic acid and sulphuric acid are good examples of bibasic acids.

$$CO\begin{matrix}OH\\OH\end{matrix} \quad \text{Carbonic acid or } H_2CO_3.$$

$$CO\begin{matrix}OH\\ONa\end{matrix} \quad \text{Sodium bicarbonate or } NaHCO_3.$$

$$CO\begin{matrix}ONa\\ONa\end{matrix} \quad \text{Sodium carbonate or } Na_2CO_3.$$

$$SO_2\begin{matrix}OH\\OH\end{matrix} \quad \text{Sulphuric acid or } H_2SO_4.$$

$$SO_2\begin{matrix}OH\\ONa\end{matrix} \quad \text{Sodium bisulphate or } NaHSO_4.$$

$$SO_2\begin{matrix}ONa\\ONa\end{matrix} \quad \text{Sodium sulphate or } Na_2SO_4.$$

A **tribasic** acid contains three atoms of replaceable hydrogen, and accordingly can form salts containing one, two or three atoms of metal. Phosphoric and arsenic acids are good examples.

$$PO\begin{matrix}OH\\OH\\OH\end{matrix} \quad \text{Orthophosphoric acid or } H_3PO_4.$$

$$PO\begin{matrix}OH\\OH\\ONa\end{matrix} \quad \text{Monosodium phosphate or } H_2NaPO_4.$$

$$PO\begin{matrix}OH\\ONa\\ONa\end{matrix} \quad \text{Disodium phosphate or } HNa_2PO_4.$$

$$PO\begin{matrix}ONa\\ONa\\ONa\end{matrix} \quad \text{Trisodium phosphate or } Na_3PO_4.$$

Calculation of formulæ.

The analysis of any compound can only give the weight of the several parts of which it is made up: the results are usually calculated into percentages, from which we have to deduce a formula for the compound. For example, an analysis of a silver salt gave the following percentage results:—

$$\begin{array}{lll} \text{Silver} & \text{Ag} = & 70.13 \text{ per cent.} \\ \text{Nitrogen} & \text{N} = & 9.09 \text{ ,,} \\ \text{Oxygen} & \text{O} = & \underline{20.78} \text{ ,,} \\ & & 100 \end{array}$$

If x be the number of atoms of silver (atomic weight $= 108$), and M the molecular weight of the compound, it is obvious that

$$100 : 70.13 :: M : 108 \times x; \quad \therefore \quad x = \frac{70.13 \times M}{108 \times 100}.$$

So if y atoms of nitrogen (atomic weight $= 14$) are present,

$$y = \frac{9.09 \times M}{14 \times 100},$$

and z atoms of oxygen (atomic weight $= 16$),

$$z = \frac{20.78 \times M}{16 \times 100}.$$

Hence we get in each case the percentage divided by the atomic weight of the elements, $\times \frac{M}{100}$ in which M is unknown. By making the division we get

$$\begin{array}{llll} \text{Ag} & 70.13 \div 108 & = & .649 \text{ or } 1. \\ \text{N} & 9.09 \div 14 & = & .649 \text{ or } 1. \\ \text{O} & 20.78 \div 16 & = & 1.298 \text{ or } 2. \end{array}$$

We now **divide** each by $.649$ (which is $\frac{100}{M}$ or some simple multiple), and the ratios obtained are $1 : 1 : 2$, and the simplest formula must be $AgNO_2$.

And if $.649 = \frac{100}{M}$ it follows that the molecular weight

$$M = \frac{100}{.649} = 154.$$

CHAPTER XVI.

METALS. POTASSIUM, SODIUM.

The metallic elements are distinguished from non-metallic elements partly by physical and partly by chemical characters, but it is difficult to draw an absolute line of separation. There are certain properties however which most members of the class metals exhibit, of which the foremost is metallic lustre: the clean surface of many metals is brilliant and reflects light. They are mostly fusible, the temperature of fusion varying from $-40°$ in the case of mercury to the intense heat of the oxyhydrogen blowpipe which is required to melt the platinum metals.

The fused metals by intermixture or chemical combination form **alloys**, which also have the metallic lustre and appearance.

Several metals are ductile, malleable, tenacious and of great tensile strength, and on account of these properties are of great use to mankind; among metals too are found the best conductors of heat and electricity.

As regards chemical functions the metals have generally an opposite polarity to the non-metals. While the oxides of many non-metals form acids, the oxides of metals form bases with opposite characters, which neutralise the former in combining to produce salts. In the electrolysis of metallic salts the metals in the bases appear at the negative pole, the non-metals at

Alkali Metals.

the positive pole, so that as a class the metals are electro-positive in relation to the non-metallic elements. The monad and dyad metals are the most electro-positive of the class and the most powerfully basic, but the pentad, hexad and heptad metals furnish acid oxides of marked character, forming well-defined salts.

In the table of elements in chapter XV the metals Lithium, Sodium, Potassium, Rubidium and Cæsium are found in the first vertical row. These elements form a well-marked natural group. The metals are easily fusible and volatile, and give, either alone or in compounds, when brought into a flame, characteristic colours with distinct and definite spectroscopic lines. The hydrates are caustic alkaline bodies, with strongly basic characters, forming salts which generally are soluble in water. The metals themselves cannot be kept in the air on account of their attraction for oxygen; they decompose water with energy, forming alkaline solutions. A large number of salts formed from these elements are identical or isomorphous in their crystalline forms.

ALKALI METALS.

	At. wt.	Hydrates.	Chlorides.	Oxides.
Lithium	Li 7	LiOH	LiCl	Li_2O.
Sodium	Na 23	NaOH	NaCl	Na_2O.
Potassium	K 39	KOH	KCl	K_2O.
Rubidium	Rb 85	RbOH	RbCl	Rb_2O.
Cæsium	Cs 133	CsOH	CsCl	Cs_2O.

Potassium (Kalium) K = 39.

The metal is never found free in nature, but is present as silicate, etc., in the older plutonic rocks (granite, etc.), and in small quantity in most soils. Fresh waters contain small but varying amounts of potassium salts, but in sea-water the element is present in large quantity, chiefly as chloride. In the ashes of plants and animals potassium as carbonate, chloride, etc., is invariably present.

The most abundant supply of potassium salts is found in the enormous salt deposits of Stassfurt, near Magdeburg, where the chloride (Silvine) KCl and Carnallite ($KCl + MgCl_2 + 6H_2O$) are largely worked.

Plant ashes, especially wood ashes, the ash of seaweeds, and the refuse from the manufacture of sugar from cane or beet-root, when burnt, furnish potassium carbonate. The residuum from grapes in wine-making also contains potash. If we burn a splinter of wood and place the ash on a moistened turmeric paper we shall find it has an alkaline reaction: this is the oldest method for obtaining potash. Waste wood and twigs are burned in heaps until a quantity of wood ashes is obtained; these are then *lixiviated* or extracted with water and the lye evaporated. A crude potassium carbonate is left (crude potashes), which after purification is known as *pearl ash*.

Metallic potassium was first obtained by Sir Humphrey Davy in 1807 by the decomposition of potash by means of a powerful current of electricity, but it is now made from the carbonate by heating it with carbon.

A mixture of potassium carbonate with charcoal is placed in an iron retort and brought to an intense red heat in a furnace. The liberated metal distils over as a greenish vapour, and is condensed in flat iron receivers dipping into naphtha.

$$K_2CO_3 + 3C = K_2 + 3CO.$$

The preparation of the metal is a difficult and somewhat dangerous operation owing to the tendency of carbonic oxide to unite with it, thereby forming an explosive substance.

Potassium is a brilliant silver white metal, soft enough to be cut with a knife, lighter than water (sp. gr. ·875), very fusible (melting point 62°·5), and easily volatilized.

If we heat a piece of potassium in a small iron spoon it melts easily and burns in the air with a lavender flame, or if a fragment is heated in a closed glass tube we may notice the formation of a green-coloured vapour. A clean piece of the metal sealed up in a tube (which has been exhausted of air) can be melted, and will adhere to the glass, showing its brilliant white metallic surface.

Metallic potassium is very easily oxidised in the air, and must therefore be preserved under naphtha or some hydrocarbon, which is without action on it. When a piece of the metal is thrown on water it at once decomposes the water, liberating hydrogen and dissolving as the hydrate (KOH) or caustic potash: the liquid tested with turmeric paper will be found to be alkaline.

$$K_2 + H_2O = 2 KHO + H_2.$$

The following are some of the chief salts of this metal:—

KCl	potassium chloride.	KOH	potassium	hydrate.
KBr	,, bromide.	KNO$_3$	,,	nitrate.
KI	,, iodide.	K$_2$CO$_3$	,,	carbonate.
KCN	,, cyanide.	KHCO$_3$	,,	bicarbonate.

Potassium hydrate, or **caustic potash**, **KOH**, is formed when the metal dissolves in water, but in practice it is prepared from the carbonate.

The carbonate is dissolved in water, slaked lime added and the mixture boiled; calcium carbonate is formed which precipitates while caustic potash remains in solution.

$$K_2CO_3 + Ca(OH)_2 = 2 KOH + CaCO_3.$$

The solution of carbonate of potash must be sufficiently diluted or the change does not take place. The clear solution is boiled down (without allowing carbonic acid of the air to have access to it) and concentrated and the heating is continued until almost a red heat is reached. The fused potash is then commonly cast into sticks by being poured into iron moulds.

Potassium hydrate is a hard, white, semicrystalline solid, liquefying quickly in the air from absorption of moisture, and also greedy of carbonic acid. It is a corrosive caustic substance, and destroys animal and vegetable tissues. Potash is very soluble in water with development of heat, and the solution has a strongly alkaline reaction, with a soapy feeling between the fingers: it also dissolves in alcohol.

When potash is neutralized with acids a salt is formed; thus, if the solution be mixed with dilute sulphuric acid until the liquid is without action on litmus paper, it will contain potassium sulphate, which can be obtained in crystals by evaporation.

$$2KOH + H_2SO_4 = K_2SO_4 + 2H_2O.$$

Potash and sulphuric acid yield potassium sulphate and water.

One characteristic property of alkalies is their action with metallic salts, which are usually decomposed with separation of the metallic hydrate, as a precipitate. Thus to a solution of copper sulphate add a little potash solution; a blue precipitate of copper hydrate will form—

$$CuSO_4 + 2KOH = K_2SO_4 + Cu(OH)_2.$$

Copper sulphate with potassium hydrate forms potassium sulphate and copper hydrate. This reaction is typical of the behaviour of alkalies with solutions of the heavy metals.

Another important character of alkalies is their power

of converting fats or oils into **soaps**, which is a change similar in kind to the decomposition of metallic salts. Boil some strong potash with a little oil for a short time, the oil will dissolve and become completely saponified. The potash, by combining with the fatty acids, stearic and oleic acids chiefly, forms potassium stearate and oleate or 'soft soap,' and glycerine, $C H_5(OH)_3$, is set free; with soda in a like manner hard soaps are produced.

Potassium forms two carbonates, neutral carbonate K_2CO_3, and bicarbonate $KHCO_3$.

Potassium carbonate, K_2CO_3, or pearl ash, is obtained by lixiviating wood ashes, or by calcining the crude tartar of wine,—potassium bitartrate, and washing the residue. It is strongly alkaline, very deliquescent and soluble. When a stream of carbonic acid is passed over it the bicarbonate is formed.

Potassium nitrate, KNO_3: nitre or saltpetre is formed in soils by the natural process of nitrification already described. It is useful as a source of nitric acid or as an oxidizing agent: very large quantities are employed for making gunpowder.

An English powder was found to contain:—

> Charcoal 13.7 per cent.
> Sulphur 10.1 ,,
> Nitre 76.2 ,,

Potassium forms three oxides, K_2O; K_2O_2; K_2O_4: they are all obtained by the direct combination of the metal with dry oxygen.

Sodium [Natrium]. Na=23.

The element sodium is abundant and widespread, entering as it does into the composition of an enormous number of rocks, and being present in traces everywhere.

Large deposits of the chloride are found as rock salt and of the nitrate as Chili saltpetre, while the carbonate, borate and sulphate are met with in lesser quantities. Sodium is present in natural waters, salt springs, and in inexhaustible quantity in sea-water. As a constituent of plants it is always found in the ash, although not so abundantly as potassium, with however the exception of a few plants inhabitants of salt marshes, which contain sodium in very large quantity. The metal was discovered by Sir Humphrey Davy, and obtained by electrolysis of the hydrate in the same manner as potassium. It may be made by decomposing caustic soda ($NaOH$) with metallic iron containing carbon, but is more usually made from the carbonate by heating it with charcoal. The process is like that described for making potassium, but is more easily managed.

Sodium is a silver white brilliant metal, with a very slight pink tinge; it is soft, of density ·973, fusible at 96°, and volatile at red heat. When thrown on water it swims about liberating hydrogen and forming solution of caustic soda: in the air it oxidizes, forming first oxide, then hydrate and carbonate.

When sodium and potassium are melted together in molecular proportions, they form a liquid alloy of brilliant metallic appearance, which may be preserved in a sealed vacuous tube.

The chief compounds of sodium are:—

$NaOH$	Sodium	hydrate.	Na_2HPO_4	Sodium	phosphate.
$NaCl$	,,	chloride.	Na_2SO_4	,,	sulphate.
$NaNO_3$	,,	nitrate.	$Na_2S_2O_3$	,,	thiosulphate.
Na_2CO_3	,,	carbonate.			

Sodium chloride is the most abundant of the sodium compounds: it is found as **rock-salt** in deposits of different geological ages, but probably all formed by drying up of

Alkali Manufacture.

salt lakes or inland seas. In Cheshire are found extensive beds; at Stassfurt, as already mentioned; at Salzburg in the Tyrol, and at various other localities in many parts of the world rock-salt deposits are worked.

The salt is either dug out solid or it may be pumped up as brine, produced by letting in water, which soon takes up as much salt as it can dissolve. The salt is recovered by evaporation. A considerable amount of salt is extracted from sea-water which contains about 3·5 per cent. of solid matters, rather less than 3 per cent. being salt. The evaporation of the water is mainly effected by natural agencies—sun and wind; artificial heat being only used in the final stages.

On the low flat coast of Hayling Island, near Portsmouth, are **salterns** which have been worked for centuries. The sea-water is taken into shallow pans where in about a week it becomes brine, which is then concentrated by boiling until the salt deposits. The residual liquors, termed bitterns, are utilised as a source of magnesia, potash, and bromine.

Sodium chloride has a well-known saline taste; it crystallises in cubes which are anhydrous. It dissolves in water, but not in alcohol; from the aqueous solution the salt can be all precipitated by hydrochloric acid gas, by which method pure sodium chloride may be prepared for laboratory use.

The Alkali Manufacture.

Common salt is the raw material from which commercial soda products are obtained, viz.:—Salt cake, Soda-ash, Carbonated and Caustic alkali.

The process known as the Leblanc process consists of several stages:—1st, production of **salt cake** by sulphuric acid; 2nd, production of **black ash**; 3rd, washing of black ash and separation of the **carbonated** and **caustic soda**.

I. **Salt cake process.** This process has for its object the production of sodium sulphate from the chloride. A charge of three quarters of a ton of salt is placed in a large cast iron pan, heated from below, and the proper quantity of sulphuric acid allowed to run in upon it. Hydrogen chloride is given off in enormous quantity and is conveyed by flues to condensing towers. The action is—

$$2NaCl + H_2SO_4 = NaCl + NaHSO_4 + HCl,$$

acid sodium sulphate and some sodium chloride being left.

The towers for condensing the hydrochloric acid are built of sandstone, and filled with lumps of coke, down which a stream of water is caused to run. The gas enters below and meets the water in its ascent, and is practically all absorbed in passing through the towers, strong solution of hydrochloric acid being drawn off at the bottom.

When the decomposition of one-half of the salt in the pan is finished, the mass left is raked into reverberating furnaces, built conveniently alongside, where it is brought to a red heat, and the action completed. This second stage is—

$$NaCl + NaHSO_4 = Na_2SO_4 + HCl.$$

Salt cake or neutral sodium sulphate, is formed, and the rest of the chlorine expelled as hydrogen chloride passes to the acid condensing towers.

II. **Black ash process.** By this process the sulphate is reduced to sulphide and converted into carbonate.

The salt cake is mixed with coal and **limestone**, $CaCO_3$, and brought into a furnace of a special kind. Formerly

Alkali Manufacture.

the reduction was performed in a reverberatory furnace, with a flat hearth, and the mixture kept stirred by workmen to expose every part to the action of the flame; but the operation is now carried on in a furnace with a revolving hearth, so that the stirring is done mechanically. The general arrangement of a revolving furnace is as follows. The fire is made in a fixed furnace at one end, from which the flame passes lengthwise through the 'revolver,' which is a large horizontal cylinder or drum sixteen to eighteen feet long. This drum is carried on strong cog-wheels, which, by acting upon a toothed girdle around it, cause it to rotate about its axis, and so the charge inside is kept turning over so as to bring it all into contact with the flame. The charge put into this furnace consists of about 30 cwt. of salt cake, 20 cwt. of coal, 30 cwt. of limestone—about 4 tons in all. The chemical changes which occur in this stage are in the main represented by two equations :—

$$\text{i. } Na_2SO_4 + C_4 = Na_2S + 4CO.$$

Sodium sulphate with carbon forms sodium sulphide and carbonic oxide.

$$\text{ii. } Na_2S + CaCO_3 = Na_2CO_3 + CaS.$$

Sodium sulphide with calcium carbonate forms sodium carbonate and calcium sulphide.

In some works quicklime is used together with the limestone, by which modification a better product is obtained.

The black ash contains approximately

Sodium carbonate	40 per cent.
Calcium sulphide	30 ,,
Lime (CaO)	10 ,,
Coal, limestone, and impurities	20 ,,

There is thus about 20 per cent. of a mixture of

unburnt coal, excess of limestone, some undecomposed salt cake, and sodium chloride, besides silica, iron, etc.

III. **Lixiviation of the Black ash.** The soluble matters in the black ash are next washed out by water. The ash is placed in tanks, conveniently arranged so that the liquid can be run from one to the other in succession, and it is washed first with strong liquors, and then with weaker liquors, and lastly with pure water, the object being to employ the least possible quantity of water, and to dissolve in it the greatest possible quantity of soda. In this lixiviation a quantity of caustic soda is produced by the lime (CaO) in the black ash acting on the sodium carbonate.

The saturated liquids are next evaporated until crystals of sodium carbonate are formed in them, which being scooped out, dried, and calcined, yield '**soda ash**' (much employed by *soap*-makers and *glass*-makers), and by solution and recrystallisation are converted into '**soda crystals**' (Na_2CO_3, $10H_2O$), the finished product. Further from the mother liquors of the black ash **caustic soda** is obtained.

A process for the direct conversion of sodium chloride into carbonate is now largely coming into use instead of the older method. It is called the Ammonia-Soda process, and depends upon the action of ammonium carbonate upon sodium chloride:—

$$NaCl + NH_3 + H_2O + CO_2 = NaHCO_3 + NH_4Cl.$$

A solution of salt in strong ammonia is brought into an iron vessel, and carbon dioxide forced in: sodium bicarbonate being but slightly soluble is precipitated, and sal-ammoniac left in solution. The ammonia is recovered from the sal-ammoniac by acting on it with lime (or magnesia), and used for a fresh operation.

Sodium Hydrate. NaOH. Caustic Soda. This body is obtained when sodium acts on water, but is manufactured from black ash liquors. In order to obtain sodium hydrate from the carbonate in the liquors, it is necessary to add lime:—

$$Na_2CO_3 + Ca(OH)_2 = 2NaOH + CaCO_3.$$

The liquid is concentrated and the residue heated until all the water is driven off.

Sodium hydrate is a powerful caustic alkali, and generally resembles potash in chemical characters. It is used in the arts, for soap making, paper making, and generally for scouring and cleansing purposes.

Sodium Carbonate, Na_2CO_3, in the anhydrous state, is a dry white powder used in the laboratory as a flux for decomposing silicates, etc. For many purposes the double salt potassium and sodium carbonate, $KNaCO_3$, is preferred, because it is fusible at a lower temperature. When crystallized from water sodium carbonate contains $Na_2CO_3, 10H_2O$: it is used for household washing purposes under the name of *soda*.

Sodium Bicarbonate, $NaHCO_3$, is formed from the crystallised carbonate by exposure to an atmosphere of carbonic acid gas.

Sodium Sulphate, Na_2SO_4, in the anhydrous form of salt cake, is largely used by glass-manufacturers. It is soluble in water, and crystallises in large transparent crystals ($Na_2SO_4, 10H_2O$), known as Glauber's Salts.

CHAPTER XVII.

AMMONIUM SALTS.

Gaseous ammonia unites, as we have seen, directly with acids producing crystalline solids; and these compounds are salts having so close a resemblance to those of the alkali metals, that they are frequently spoken of as **ammonium salts**, and are supposed to contain the group NH_4,—which in many respects resembles an alkali metal. This may be illustrated by the following examples:—

$NH_3 + HCl \ \ = NH_4, Cl$ ammonium chloride.
$NH_3 + H_2O \ \ = NH_4, OH$,,   hydrate.
$NH_3 + HNO_3 = NH_4, NO_3$,,   nitrate.
$NH_3 + H_2S \ \ = NH_4, HS$,,   hydrosulphide.

Ammonium hydrate is an alkaline liquid, resembling solution of potash or soda; the crystallised salts are frequently similar to or identical in form (isomorphous) with those of potassium. Ammonium chloride crystallises in cubes, ammonium nitrate in prisms like those of potassium nitrate, and ammonium alum is in crystalline form identical with potash alum.

Ammonium Chloride. Sal-ammoniac. $NH_4Cl.$

This salt is made from the ammoniacal liquors of gas works. They are usually separated from tarry impurities by distillation, and neutralised with commercial hydrochloric acid: the crude salt can be purified by sublimation. Sometimes ammonium sulphate is first prepared from the gas liquors and converted into sublimed ammonium chloride by being heated with common salt.

$$(NH_4)_2SO_4 + 2NaCl = Na_2SO_4 + 2NH_4Cl.$$

The compound is also formed by the direct union of gaseous ammonia with gaseous hydrogen chloride.

Ammonium chloride is a white fibrous solid in the commercial sublimed form, but by evaporation of the solution may be obtained in cubic crystals. It dissolves in water, with a lowering of temperature, and volatilises when heated without previous fusion. The vapour density of ammonium chloride is found to be 13·37, or the exact mean of its constituent gases. The molecular weight is 53·5—

$$[NH_3 + HCl = 17 + 36.5 = 53.5]$$

and the vapour density is **one-fourth** of this value. Hence, it appears that in the state of vapour the salt is resolved into its constituents or dissociated, and the vapour is a mixture of equal volumes of hydrogen chloride and ammonia—

$$\boxed{NH_4Cl} = \boxed{HCl} + \boxed{NH_3}$$

The supposition that this dissociation really takes place has been proved to be correct by direct experiment, and free ammonia and hydrogen chloride can be detected in the vapour of sal-ammoniac by their effects on litmus paper.

Ammonium chloride, like potassium chloride, forms a yellow crystalline salt with platinum tetrachloride; the formulæ of these compounds respectively are—

$$(NH_4)_2PtCl_6 \text{ and } K_2PtCl_6.$$

An interesting experiment with this salt is its conversion into the so-called **ammonium amalgam**. A strong solution of ammonium chloride is poured upon some sodium amalgam previously formed by dissolving sodium in mercury. The amalgam swells up to a bulky pasty metallic mass, which soon, however, decomposes, giving off free ammonia and hydrogen gas.

Ammonium Sulphate, $(NH_4)_2SO_3$, is obtained from ammoniacal gas liquors by using sulphuric acid for neutralising them. It is frequently mixed with other fertilising substances in artificial manures.

Ammonium Carbonate, $(NH_4)_2CO_3$.

True normal carbonate is difficult to obtain, and difficult to keep, owing to the rapidity with which it loses ammonia; it may be formed by treating ordinary carbonate of ammonia with strong aqueous ammonia.

Commercial carbonate of ammonia—sal volatile or smelling salts—is a white solid substance made by heating together sal-ammoniac and chalk. It is a compound body containing ammonium bicarbonate, NH_4, H, CO_3, and ammonium carbamate, NH_4, NH_2CO_2. This last salt is formed by the union of two molecules of dry ammonia with carbon dioxide $(2NH_3, CO_2)$, and by addition of water becomes converted into bicarbonate.

Ammonium Hydrate, or solution of ammonia, has been described. In its strong alkaline characters it resembles caustic potash.

Ammonium Sulphide is much used as a reagent for laboratory purposes. It is made by passing hydrogen sulphide into ammonia—

$$NH_3 + H_2S = NH_4HS.$$

The liquid which contains ammonium hydrosulphide is colourless when first made, but turns yellow by keeping, and then contains polysulphides of ammonium: this yellow solution treated with acids gives off hydrogen sulphide, and deposits sulphur in the form of a white amorphous powder.

Ammonium salts are all decomposed by heating with the fixed alkalies, potash and soda, or the alkaline earths, lime, baryta, etc.; gaseous ammonia being evolved which may be recognised by its odour and behaviour with turmeric paper.

CHAPTER XVIII.

DYAD METALS. THE ALKALINE EARTHS.

The dyad metals, although forming a single series of elements in the 'periodic' arrangement, yet fall into two distinct groups, of which the metals magnesium and calcium may be taken as types. The following are the members of the two groups, with their atomic weights:—

Beryllium	Be	9	Calcium	Ca	40
Magnesium	Mg	24	Strontium	Sr	87
Zinc	Zn	65	Barium	Ba	137
Cadmium	Cd	112			

The oxides of the dyad metals are earthy powders, and, with the exception of cadmium oxide, white at ordinary temperatures; they are generally infusible and non-volatile. But the oxides of the calcium group have a great affinity for water, and are, to a certain extent, soluble in it, forming **alkaline** solutions, and these solutions of soluble hydrates will precipitate the insoluble hydrates of the metals of the magnesium group from their dissolved salts. The metals of the calcium group alone can form definite peroxides.

Magnesium and the metals allied with it are permanent in air, not forming oxides at ordinary temperatures; but the metals of the calcium group attract oxygen eagerly from the air, and can only be kept under rock oil or in an inert gas. In respect of the solubility of the sulphates very considerable differences

Calcium.

exist, for, while sulphates of magnesium, zinc, and cadmium are freely soluble in water, those of calcium, strontium, and barium are dissolved only sparingly.

Alkaline Earths.

The following are the chief compounds formed by the metals of this group:—

CaO	Lime.		SrO	Strontia.	
$Ca(OH)_2$	Calcium hydrate.		$Sr(OH)_2$	Strontium hydrate.	
CaO_2	„	peroxide.	SrO_2	„	peroxide.
$CaCl_2$	„	chloride.	$SrCl_2$	„	chloride.
$CaSO_4$	„	sulphate.	$SrSO_4$	„	sulphate.
CaN_2O_6	„	nitrate.	SrN_2O_6	„	nitrate.
BaO	Baryta.				
$Ba(OH)_2$	Barium hydrate.				
BaO_2	„	peroxide.			
$BaCl_2$	„	chloride.			
$BaSO_4$	„	sulphate.			
BaN_2O_6	„	nitrate.			

Calcium, Ca. Atomic weight, 40.

Calcium is by far the most abundant metal of the group, and in the form of carbonate—chalk and the various limestones—forms enormous rock masses in different parts of the world. In some districts the sulphate—gypsum—is found in considerable quantities. The chief native varieties of calcium carbonate, or carbonate of lime, $CaCO_3$, are (i) the amorphous forms chalk and various kinds of limestone; (ii) the crystalline-forms of marble, and (iii) the well-crystallised forms of calcite, or Iceland Spar, and Aragonite. Fluor spar, CaF_2, is found as a mineral; it crystallises in cubes.

Metallic calcium may be obtained by electrolysis of the fused chloride, or the action of metallic sodium on the chloride. The metals strontium and barium may be obtained by similar methods. They readily oxidise in the air, burning brilliantly if heated, and are commonly preserved in petroleum.

Calcium Carbonate. This compound is produced artificially by precipitation of a calcium salt with a soluble carbonate, e. g.—

$$CaCl_2 + Na_2CO_3 = CaCO_3 + 2NaCl.$$

Calcium chloride with sodium carbonate forms calcium carbonate and sodium chloride.

It is also precipitated by passing carbonic acid gas into lime water. Carbonate of lime is only dissolved to a minute extent by pure water, but water containing carbonic acid is able to dissolve it in considerable quantity. This may easily be shown. Dilute some lime water by adding an equal quantity of *distilled* water, and pass a stream of carbonic acid gas into the liquid until it becomes milky; the white precipitate which separates is calcium carbonate. Let the passing of the gas continue for a time, and the precipitate will gradually redissolve and the liquid become clear as at the beginning. The change is due to the formation of a soluble calcium bicarbonate and if the clear solution be divided into two parts, we shall find on boiling one portion that the liquid becomes milky again, carbonic acid is driven off, and the insoluble carbonate precipitates. If clear lime-water be added to the second portion, a precipitate is also produced, for the lime-water neutralising the carbonic acid, causes an entire precipitation of the calcium carbonate in solution. We may represent these operations by the following equations:—

i. $Ca(OH)_2 + CO_2 = CaCO_3 + H_2O.$
ii. $Ca(OH)_2 + 2CO_2 = CaH_2C_2O_6.$

In the first case insoluble carbonate is formed, and in the second soluble bicarbonate—

iii. $CaH_2C_2O_6 = CaCO_3 + CO_2 + H_2O.$
iv. $CaH_2C_2O_6 + Ca(OH)_2 = 2CaCO_3 + 2H_2O.$

Hardness of Water. 175

The soluble bicarbonate is converted into the insoluble carbonate by (iii) expulsion of carbonic acid, or (iv) by neutralising with lime.

These facts have an important practical bearing in relation to the **hardness** of water. In limestone districts water is what is termed **hard** owing to the difficulty with which it forms a lather with soap, and the hardness is caused by the presence of salts of lime (and possibly also of magnesia) in the water. When a soluble soda soap is used with such waters an insoluble lime soap is formed; the soluble sodium stearate being converted into calcium stearate, which curdles and separates, and a lather will not form until all the lime has been thus removed. Another difficulty with hard waters arises from the deposit of **crust** or **fur** in steam boilers and kettles; the incrustation being caused by the deposition of calcium carbonate when the water is heated, carbonic acid being expelled. For the purpose of **softening** water in steam boilers, caustic soda and alkaline salts may be used, but such bodies are not admissible in drinking waters. Water for domestic use, and especially chalk waters, may be successfully softened on a large scale by the addition of lime.

Clarke's Soap Test. In order to estimate the degree of hardness in any water a standard solution of soap is used. A measured quantity of the water—50 cc.—is poured into a bottle, and the soap solution added from a graduated burette, the bottle being vigorously shaken after each addition. After a time a lather is produced, and when the soap bubbles remain unbroken for some minutes, the quantity of soap solution which has been used is read off and the value thus obtained indicates the degree of hardness of the water.

The soap solution itself is previously standardised

by water artificially prepared with a known amount of lime in solution.

Lime. CaO. Calcium Oxide. The practice of burning limestone to make lime is one of great antiquity. When calcium carbonate is strongly heated, carbon dioxide is driven off, leaving *quick lime*. A high temperature is required, and a certain amount of moisture in the air is necessary to the success of the operation. For chemical uses marble is sometimes heated in a crucible in a furnace to furnish a purer form of lime.

Lime when exposed to the air absorbs water and carbonic acid, falling into a bulky white powder. If water is poured upon lumps of *freshly-burnt* lime, much heat is developed and the lumps swell greatly, becoming converted into a powder of slaked lime or calcium hydrate, $Ca(OH)_2$.

The expansive force produced by the swelling of lime when wetted is now utilised in getting coal when it is desirable to avoid using gunpowder; cartridges of quicklime placed in boreholes when wetted expand, and so dislocate large masses of coal.

Calcium hydrate mixed with water forms a creamy liquid—milk of lime—which becomes clear on standing. The clear lime-water contains about one-seven hundredth part of dissolved lime, and is feebly alkaline and caustic.

Slaked lime mixed with sand is used as a **mortar** for building purposes. **Hydraulic** mortars are either made from limestones (such as Blue Lias Stone), which contain an admixture of clay, or from artificial mixtures of chalk and clay. They have the property of setting hard under water.

Calcium Chloride. $CaCl_2$. This compound is obtained by dissolving the carbonate in hydrochloric acid and evaporating the liquid. If dried at a strong heat it forms a white porous mass; but if the evaporation is stopped at the right point, crystals of $CaCl_2, 6H_2O$ can be obtained. It is a deliquescent salt, absorbing water from the air and liquefying; it is very soluble both in water and in alcohol. The anhydrous salt, on account of its attraction for water, is much used for drying gases, etc. It fuses at a red heat, and the fused substance, by the action of a powerful electric current, splits up into calcium and chlorine.

Calcium Sulphate. $CaSO_4$. The native forms of this compound containing water are gypsum, alabaster, and crystals of selenite ($CaSO_4, 2H_2O$). It is found without water as anhydrite $CaSO_4$. Calcium sulphate is a common constituent in natural waters. It is only moderately soluble in water, and is thrown down by adding sulphuric acid or a sulphate to a fairly strong solution of the chloride.

Gypsum is employed for the manufacture of 'Plaster of Paris.' When heated to about 250° C. water is given off and a white substance left, which has the property of re-combining with water to form a hard mass. The act of hydration is accompanied by a rise of temperature.

Calcium salts, when volatilised in the flame of a Bunsen burner, produce a brick-red coloration.

Strontium. Sr. Atomic weight, 87.

Strontium is not very abundant, although local deposits in quantity are met with. The chief strontium minerals are celestine, $SrSO_4$, and strontianite, $SrCO_3$; the metal is frequently present in both calcium and barium minerals.

Strontia. SrO. This oxide is obtained by strongly heating the nitrate (not from the carbonate). It is caustic and alkaline, and soluble in water.

Strontium Carbonate, $SrCO_3$, is formed by precipitation of strontium chloride with an alkaline carbonate.

Strontium Nitrate. SrN_2O_6, is prepared by dissolving the carbonate in nitric acid; the salt is chiefly used for making **red fire**.

Strontium Sulphate $SrSO_4$, is formed as an insoluble precipitate upon the addition of sulphuric acid or a sulphate to solution of the chloride or nitrate. It is less soluble than calcium sulphate, but more so than barium sulphate. Most of the compounds of strontium impart a bright crimson tint to the flame of a Bunsen burner. In chemical properties the salts of strontium are intermediate between those of the allied metals calcium and barium.

Barium. Ba. Atomic weight, 137.

The chief mineral forms in which barium is found are Heavy Spar, $Ba\,SO_4$, and Witherite, $Ba\,CO_3$. To obtain barium chloride, the sulphate is mixed with carbon and heated, whereby a reduction to sulphide takes place:—

$$BaSO_4 + C_4 = BaS + 4CO,$$

and by treatment of the sulphide with dilute hydrochloric acid a solution of chloride is obtained; or otherwise, a mixture of heavy spar with carbon and calcium chloride is heated, the barium being at once obtained as chloride, and the calcium as sulphide; the two products are separated by washing, when the barium chloride dissolves out.

Barium Chloride. $Ba\,Cl_2$, is a white anhydrous salt, soluble in water, but differs from calcium chloride in not being deliquescent: it is useful as a laboratory reagent.

Barium Nitrate, Ba N_2O_6, is most easily obtained by dissolving Witherite (barium carbonate) in dilute nitric acid, and evaporating the solution till crystals form. It is soluble in water, but not deliquescent. This salt is used in the preparation of **green fire**.

Barium Oxide or Baryta. Ba O. This oxide is obtained, like strontia, by strongly heating the nitrate, when it is left as a greyish porous mass. It combines with water and carbonic acid with more avidity even than lime, and is more soluble than the latter in water. Baryta water is a caustic alkaline liquid; it absorbs carbonic acid readily, and may be used as a test for that gas; when evaporated the solution deposits crystals of barium hydrate, BaO_2H_2, $8H_2O$.

Caustic Baryta is obtained for commercial purposes by an interesting process. Barium sulphide is prepared from heavy spar as above described, and the sulphide is converted into carbonate by contact with carbonic acid. The carbonate is then subjected to the action of superheated steam, when the following change takes place:—

$$BaCO_3 + H_2O = Ba(OH)_2 + CO_2.$$

Barium carbonate and steam produce barium hydrate and carbon dioxide.

Barium Dioxide. BaO_2. Baryta absorbs oxygen at a red heat forming the dioxide: this is a white powder, soluble in dilute hydrochloric acid, and producing hydrogen dioxide—

$$BaO_2 + 2HCl = BaCl_2 + H_2O_2.$$

Barium dioxide, when strongly heated, gives off oxygen, leaving baryta; and a process for the commercial preparation of oxygen by this method is in successful operation. Pure baryta is heated in a retort, and dry air free from carbonic acid is pumped in so long as the oxygen is taken up, the nitrogen being allowed to escape; then, on raising the temperature, oxygen is given off, which is pumped out from the retorts into holders, and when the oxygen has been exhausted, a fresh quantity of air is pumped in and the operation can be repeated an indefinite number of times with the same baryta.

Barium Carbonate. $Ba\,CO_3$. This compound is found native as Witherite, and may be artificially prepared as a white powder by precipitating barium solutions with soluble carbonates.

Barium Sulphate. $Ba\,SO_4$. The mineral form of barium sulphate is, as its name Barytes or Heavy Spar implies, remarkable for its weight, the density being 4·5. This compound is precipitated by the addition of sulphuric acid, or any sulphate, to a solution of barium chloride, etc., and is a remarkably insoluble substance. It is almost entirely insoluble in water, and very slightly soluble in acids. Solutions of calcium sulphate, or of strontium sulphate, are employed as a test for barium, as when mixed with barium solutions they produce a precipitate of the extremely insoluble sulphate.

Barium salts impart a **green** colour to the flame of a Bunsen burner.

CHAPTER XIX.

DYAD METALS. BERYLLIUM, MAGNESIUM, ZINC, AND CADMIUM.

The metals of this group are permanent in air, but when heated burn readily and produce oxides. They are all volatile; but while magnesium is vaporised with difficulty, zinc and cadmium are distilled with ease. The following are some of the typical compounds of these elements:—

MgO	Magnesium oxide.	ZnO	Zinc oxide.
$MgCl_2$	,, chloride.	$ZnCl_2$	,, chloride.
MgN_2O_6	,, nitrate.	ZnN_2O_6	,, nitrate.
$MgSO_4, 7H_2O$	,, sulphate.	$ZnSO_4, 7H_2O$	,, sulphate.
	CdO	Cadmium oxide.	
	$CdCl_2$	,, chloride.	
	CdN_2O_6	,, nitrate.	
	$CdSO_4$	,, sulphate.	

Beryllium. Be. Atomic weight, 9·3.

The rare metal beryllium found in Beryl, $3BeO, Al_2O_3, Si_3O_6$, belongs to the dyad group of metals, and bears a general resemblance to magnesium in the character of its compounds.

Magnesium. Mg. Atomic weight, 24.
Molecular weight [?]. Specific gravity, 1·7.

Magnesium is a less abundant element than calcium, but is almost invariably found in association with it. Magnesium carbonate is found as magnesite, $MgCO_3$, but usually occurs along with calcium carbonate in dolomite and magnesian limestone. Magnesium, as silicate, enters into the composition of numerous minerals. The sulphate and carbonate may be found in many natural waters, and from the bitter mother liquors left after the evaporation of sea water both sulphate and chloride can be obtained. A double chloride of magnesium and potassium, known as Carnallite ($MgCl_2$, KCl, $6H_2O$), is worked at Stassfurt for the preparation of salts of the two metals.

The metal Magnesium is manufactured from the chloride by the action of metallic sodium at a high temperature. It can be obtained also by electrolysis of the fused chloride. It is a white, very light metal (sp. gr. 1·7), permanent in air, but slowly becoming covered with a film of oxide in moist air. It dissolves readily in dilute acids with liberation of hydrogen.

$$Mg + 2HCl = MgCl_2 + H_2.$$

A strip of ribbon of metal burns with a bright, very white flame forming the oxide, MgO.

Magnesium Sulphate, $MgSO_4$, $7H_2O$.

This compound, known also as **Epsom salts**, is obtained from the mother liquors of sea water, or by dissolving the native carbonate in sulphuric acid. It is readily soluble in water and has a bitter saline taste. A native form of the sulphate is known as Kieserite, $MgSO_4$, H_2O; it is almost as insoluble in water as gypsum, but by long contact with water slowly changes into Epsom salts, $MgSO_4$, $7H_2O$, and passes into a soluble condition.

Magnesium Carbonate, $MgCO_3$, is obtained by adding the sulphate to a cold solution of sodium carbonate: but if the solutions are hot the precipitate is basic carbonate, i.e. a mixture of hydrate and carbonate.

An ingenious process for the manufacture of magnesium carbonate was devised by Pattinson. Dolomitic limestone is gently ignited so that the carbonate of magnesia decomposes, while the carbonate of lime remains unaltered; it is then treated with carbonic acid and water under considerable pressure, and the magnesia passes abundantly into solution as bicarbonate leaving the lime undissolved. The solution thus obtained being heated by steam, deposits the dissolved magnesia in the form of a light white powder.

Magnesium Chloride, Mg Cl$_2$. A solution of the chloride is obtained by dissolving either the metal or (more economically) the carbonate or oxide in hydrochloric acid. The liquid, by slow evaporation, deposits crystals, Mg Cl$_2$, 6 H$_2$O, which on heating decompose, hydrochloric acid being expelled and magnesia left. To prepare the anhydrous chloride, a double salt must first be made by adding ammonium chloride to the solution (NH$_4$ Cl, MgCl$_2$, 6 H$_2$O): this compound if carefully heated leaves a residue of the chloride, Mg Cl$_2$, while the ammonium salt volatilises.

Magnesium Oxide, Magnesia, MgO.

The oxide is made by heating the carbonate to low red heat, when it loses all its carbon·dioxide and no longer effervesces with acids. When lime-water is added to a solution of magnesium chloride, etc., a precipitate of the hydrate is formed, Mg(OH)$_2$. Both magnesia and the hydrate are to a slight degree soluble in water, but the solution shows a very feeble alkalinity. The hydrate, carbonate, and oxalate of this metal are insoluble in pure water, but are readily soluble in presence of ammonium salts such as ammonium chloride. Magnesium is identified, in analysis, by the fact that it forms a crystalline double salt with phosphates in the presence of ammonia and ammonium chloride; the formula of the salt being $MgNH_4, PO_4, 6H_2O$.

Zinc. Zn. Specific gravity, 7·0.
Atomic weight, 65. Molecular weight, 65.

The chief ores of zinc are the sulphide or blende, ZnS, the oxide, ZnO, and the carbonate or calamine, $ZnCO_3$; and there is also a silicated calamine, Zn_2SiO_4. To obtain the metal the ores are roasted, whereby zinc oxide is formed, the sulphur, if present, burning off as sulphur dioxide. A mixture of this oxide with powdered carbon (coal or coke) is placed in a retort which can be heated strongly from the outside; the zinc oxide is reduced, and the metal being volatile passes from the retort, and is collected in receivers.

$$ZnO + C = Zn + CO.$$

The boiling-point of zinc is 940°C.; it melts at about 430°. When strongly heated in the air it takes fire, burning with a bluish-white flame and producing flakes of white oxide.

Zinc is a bluish-white crystalline metal and brittle under ordinary conditions, but it has the remarkable property of becoming malleable and workable if moderately heated; and at a temperature of 100° to 150° it can be rolled into sheets.

Clean iron brought into a bath of melted zinc becomes alloyed on the surface with an adherent coating: the product is technically known as **galvanised iron.**

Dilute sulphuric or hydrochloric acids and also potash dissolve metallic zinc with a liberation of hydrogen gas.

Zinc Oxide, ZnO. The oxide is obtained when the metal burns in air and oxygen, but is best prepared by heating the carbonate. A solution of sodium carbonate

Zinc Salts.

precipitates zinc carbonate, more or less mixed with hydrate, which, when washed, dried, and ignited, is changed into the oxide. The oxide is a white infusible solid which turns yellow when heated.

Zinc Hydrate, $Zn(OH)_2$, is precipitated by ammonia or alkalies from zinc solutions. It dissolves in potash or soda, probably forming with potash $Zn(OK)_2$.

Zinc Sulphide, Zn S. When neutral or alkaline solutions of zinc are treated with hydrogen sulphide a white precipitate of zinc sulphide (hydrated) is obtained; the precipitation is prevented by the presence of free mineral acid. The same white sulphide is formed when ammonium sulphide is added to solutions containing zinc; the reaction is a useful test for this metal.

Zinc Sulphate, $Zn SO_4, 7 H_2O$. This is a white crystalline salt, resembling Epsom salts in appearance: it has an astringent taste and, like other soluble zinc salts, is poisonous. Zinc sulphate is formed by dissolving either the metal, the oxide or carbonate, in sulphuric acid.

Zinc Chloride, $Zn Cl_2$. The chloride is formed when metallic zinc dissolves in hydrochloric acid; on evaporation of the liquid a white solid is obtained which is extremely deliquescent. A strong solution of this compound corrodes and chars organic substances; solid zinc chloride is used in surgery as a caustic; the solution, when diluted, is employed for an antiseptic or disinfectant, and is known as Sir Wm. Burnett's fluid.

Cadmium. Cd. Specific gravity, 8·6.
Atomic weight, 112. Molecular weight, 112.

Cadmium is associated in nature with zinc, and is rarely found except in zinc ores. It reduces along with the zinc in the manufacture of that metal, and being more easily volatile distils off with the first portions of the zinc; and these being kept apart are afterwards used for preparing the metal.

Cadmium is a white metal, permanent in air, and resembling zinc in many physical and chemical characters. It melts at 315°C., and at a temperature of 772° boils and passes into vapour. If a piece of metallic cadmium in a hard glass tube be heated in a current of hydrogen, the metal distils, and on cooling, bright globules with crystalline facets are obtained.

Cadmium Oxide, Cd O. The anhydrous oxide of cadmium is *brown* in colour, in this respect differing from the other oxides of the group, all of which are *white*. It may be made by igniting either the hydrate, carbonate, or nitrate. Cadmium hydrate, $Cd(OH)_2$, is white; it is precipitated from the solutions by alkalies, is soluble in ammonia, but not in potash—in this latter particular differing from zinc.

Cadmium Sulphide, Cd S. The sulphide is obtained as a **yellow** precipitate by treating cadmium solution with hydrogen sulphide. It is soluble in acids, and the precipitation is prevented by the presence of an excess of free mineral acid. Yellow cadmium sulphide is employed as a pigment.

Cadmium Sulphate, $Cd SO_4$. Most of the salts of cadmium are white in colour, and the sulphate is a white crystalline salt, containing somewhat variable quantities of water according to the conditions under which the solution is crystallised. A solution of the sulphate is made by dissolving cadmium metal, oxide, sulphide, or carbonate in sulphuric acid.

Vapour density and Molecular weight.

The specific gravities and the atomic weights of the metals magnesium, zinc, and cadmium increase together, but the series of numbers representing their fusibility and volatility is in descending order :—

	At. wt.	Sp. gr.	Melt. pt.	Boil. pt.	Vapour densy.
Magnesium	24	1·7	?500	?1100	?
Zinc	65	7·0	430	940	32·5
Cadmium	112	8·6	315	772	56

The vapour density of the metal magnesium is unknown, partly because of the high temperature at which the metal volatilises, but the vapour densities of both zinc and cadmium are well ascertained; the values being approximately 32·5 and 56.

The molecular weights, therefore, of these elements are 65 and 112, and are identical with the atomic weights.

The vapour density of an organic zinc compound :— zinc ethyl, $Zn(C_2H_5)_2$, has been found by experiment to be 61·5, which accords with the formula, i. e.

$$\frac{65 + 48 + 10}{2} = 61.5.$$

CHAPTER XX.

TRIAD METALS. ALUMINIUM.

The only members of the triad series of elements commonly met with are Boron and Aluminium: all the rest are rare metals (Gallium, Indium, Thallium).

Aluminium is the most abundant of all metals, and of the three elements of which the crust of the earth is principally composed, viz. oxygen, silicon, aluminium, it comes third in order of quantity.

Aluminium. Al. Atomic weight, 27. Specific gravity, 2·6.

The metal aluminium exists in nature chiefly as silicate: it is the metal of clay, and enters, as a silicate, into the composition of an enormous number of minerals and rocks. Pure clay or Kaolin is $Al_2 Si_2 O_7, 2 H_2 O$, and results from the decomposition of felspar from granite:—

[Felspar $K_2 Al_2 O_4, Si_6 O_{12}$ or $2 KAlSi_3 O_8$].

The varieties of clay-slates and shales consist chiefly of aluminium silicate.

The oxide $Al_2 O_3$ occurs in the extremely hard minerals or gems **corundum, emery, sapphire,** and **ruby,** or combined with magnesia as $Mg Al_2 O_4$ in **spinel. Cryolite** is a double fluoride of sodium and aluminium, $Na_3 Al F_6$.

It is very difficult to prepare aluminium from the oxide, as carbon only decomposes this body at the most intense heat obtainable. Reduction takes place when alumina is placed between the carbon poles of a powerful voltaic arc, but in practice the metal is obtained from the chloride by the action of sodium.

The metal is brilliant white in colour, takes a high polish, and does not tarnish in air; it is remarkably resonant when struck; very tough and light, and can be forged and worked into almost any shape. If it were produced at less cost it would probably come very largely into use.

Aluminium does not oxidise in air even when heated, but it may be burnt in oxygen gas. Nitric acid is almost without effect on it; but with chlorine gas it enters energetically into combination or dissolves freely in hydrochloric acid: also in potash it dissolves with liberation of hydrogen.

The following are the most important and typical compounds:—

Al_2O_3, Aluminium oxide.
$Al(OH)_3$, Aluminium hydrate.
$AlCl_3$, Aluminium chloride.
$KAl(SO_4)_2, 12H_2O$, Potash alum.
$NH_4Al(SO_4)_2, 12H_2O$, Ammonia alum.

Aluminium Oxide or **Alumina**, Al_2O_3, is obtained artificially by igniting the precipitated hydrate or sulphate. A white, porous solid, infusible, except by the oxyhydrogen blowpipe.

Aluminium Hydrate, $Al(OH)_3$, is thrown down as a white gelatinous precipitate by adding alkali to solution of aluminium salts; in excess of potash or soda it dissolves, forming $Al(OK)_3$, thus acting as a weak acid; it is soluble also in dilute acids, but its basic characters are feeble, and it will not combine with carbonic acid. The mineral diaspore is a native hydrate, H_2O, Al_2O_3.

Aluminium Chloride, Al_2Cl_6, or $AlCl_3$, is formed in a manner similar to silicon chloride. A mixture of alumina and carbon is heated to full redness, while a current of chlorine passed over it; the chloride being volatile passes forward and is condensed in a receiver. If common salt is previously added to the mixture the compound Na Al Cl$_4$ is obtained (Na Cl, Al Cl$_3$). This sodium aluminium chloride is used for preparing metallic aluminium.

Aluminium chloride is a yellowish solid body, fuming in moist air and giving off hydrogen chloride. It combines eagerly with water and decomposes probably into aluminium hydrate and hydrogen chloride thus:—

$$AlCl_3 + 3H_2O = Al(OH)_3 + 3HCl.$$

Both bodies remain in solution, but if the liquid be evaporated hydrochloric acid passes off and alumina is left. A form of soluble aluminium hydrate may be obtained by dialysis from this solution in a similar manner to that used in preparing soluble silica. Hydrated crystals of aluminium chloride are formed by slow evaporation of the solution.

Aluminium chloride boils at 180°C., and the density of the vapour at 900°C. has been recently found to be nearly 67; the molecule at high temperatures is therefore represented by the formula $AlCl_3 = 134$.

Aluminium Sulphate, $Al_2(SO_4)_3$, is sometimes found as a mineral ($Al_2(SO_4)_3$, $18H_2O$), but is difficult to obtain artificially in crystals. It is obtained pure by dissolving the hydrate in sulphuric acid; and for commercial uses as a mordant in dyeing, large quantities are manufactured by dissolving burnt clay in sulphuric acid.

Alums. The substance originally known as alum was the double sulphate of potassium and aluminium; but other double salts of similar constitution are also described as alums.

The commonest alums are :—

    Potash alum = $KAl(SO_4)_2, 12H_2O.$
    Ammonia alum = $NH_4Al(SO_4)_2, 12H_2O.$
    Iron ammonia alum = $NH_4Fe(SO_4)_2, 12H_2O.$
    Chrome alum = $KCr(SO_4)_2, 12H_2O.$

An alum is, in fact, a double sulphate, containing an alkali and a metallic sesqui-oxide, combined with twelve molecules of water and crystallised in regular octahedra (fig. 53). $K_2SO_4 + Al_2(SO_4)_3 + 24H_2O = 2(KAl(SO_4)_2, 12H_2O)$

In the manufacture of alum the first step is the preparation of crude aluminium sulphate, either from clay by the action of sulphuric acid, or by roasting bituminous shale containing pyrites (FeS_2). When the roasted shale is lixiviated the solution contains aluminium sulphate and also sulphuric acid (with some iron sulphate, etc.); it can be converted into alum by the addition of an ammonium or potassium salt in solution. The liquid is evaporated until crystals of alum are deposited.

Fig. 53.

Clay, pottery, porcelain, etc.

Clays are a combination of alumina, silica, and water for the most part, but with smaller and varying amounts of iron, lime, magnesia, and alkalies. In a moist condition clay is impervious to water, but able to hold a considerable quantity by mechanical adhesion; on drying it shrinks considerably, and if strongly ignited parts with all combined water, and forms a porous, infusible, brick-like mass.

The finest pottery made from clay is **porcelain**, which has a well-known semi-transparent, fused glassy structure. This character is produced by mixing with china clay used in the manufacture a fusible 'frit' or flux which, uniting with the porous clay when fired, converts it into the close-grained impervious porcelain. Felspar, with the addition of some quartz, is commonly used as the frit or flux. To produce a glazed surface the ware, after a preliminary drying and baking, is coated by being dipped into a thin mud of ground felspar and quartz, which is more fusible than the body of the material. A very high temperature is employed in the kilns in which the porcelain is finally burnt; the glaze fuses into a colourless, strongly adherent coating, which, when cold, is able to resist the action of strong acids and alkalies.

Ordinary earthenware is made from commoner clays, frequently mixed with chalky marl to prevent shrinkage. In many cases the glazing is done with common salt. After the pottery is thoroughly burnt, and while the temperature of the kiln is very high, a quantity of salt is thrown in which passes into vapour and acts upon the silica of the heated clay; thus a superficial coating of glass or sodium silicate is produced upon the ware.

'Stoneware' is a common sort of porcelain, as it is made of a mixture of clay with some fusible frit.

Unglazed bricks, tiles, flower pots, and the finer forms of unglazed pottery known as terra cotta are produced from clays of various kinds by a simple burning or baking process.

Fire clays are rich in silica, and produce bricks of a refractory and very infusible character.

CHAPTER XXI.

IRON. NICKEL. COBALT.

THESE three elements come next after Chromium and Manganese, in the order of atomic weights, and the atomic weights of the five elements lie close together: thus Chromium is 52, Manganese 55, Iron 56, Nickel 58·6, and Cobalt 59. In their chemical characters they have many points of similarity. The lowest oxides are of the dyad type RO, and they form sulphates and chlorides, represented by the forms $R\,SO_4$ and $R\,Cl_2$; the sesqui-oxides, R_2O_3, and the triad chlorides, RCl_3, of chromium and of iron are very definite and stable compounds, resembling aluminium oxide and chloride. Chromium and manganese, however, by reason of the formation of higher oxides of acid properties which, in combination with bases produce chromates and manganates, are separated from the others.

Iron. Ferrum. Fe. Atomic weight, 56. Specific gravity, 7·8.

Iron is a metal which is abundant in all parts of the world, but of limited occurrence in the native state. Masses of iron of meteoric origin occasionally fall upon the earth or are found near the surface: they always contain some nickel and cobalt.

The chief ores of iron are the sesqui-oxide or hæmatite Fe_2O_3, and magnetic oxide Fe_3O_4; a hydrated oxide Limonite $2Fe_2O_3, 3H_2O$ is also used as an ore. In some districts varieties of ferrous carbonate or spathic iron ore $Fe\,CO_3$, are smelted. The sulphide, pyrites. FeS_2, is not worked as iron ore.

In addition to the particular accumulations of iron used for the extraction of the metal, small quantities of this element are found in numerous rocks and minerals.

The reduction of oxide of iron to the metallic state is very easy, and can be shown by the following experiment:—

Place a small quantity of dry ferric oxide in a hard glass tube and pass a stream of dried hydrogen over it. When the air is expelled from the apparatus apply a Bunsen burner, so as to heat the oxide to redness, when the reduction will take place (fig. 14). After the lamp is removed the metal should be cooled in the current of gas and examined when cold. It will be a gray, metallic powder, attracted by the magnet, and soluble in sulphuric acid with evolution of hydrogen. The reduction is represented by the equation—

$$Fe_2O_3 + 3H_2 = Fe_2 + 3H_2O.$$

A similar reduction is effected by carbonic oxide thus—

$$Fe_2O_3 + 3CO = Fe_2 + 3CO_2.$$

Production of Cast Iron.

The operation of Iron Smelting is carried out on a very large scale in the blast furnace; a structure of 60 to 90 feet high, which is shown in section in the drawing, fig. 54.

Ironstone and coal and limestone are successively thrown in at the top of the furnace, and a powerful blast of air is driven in through special openings beneath, so as to maintain a state of intense and rapid combustion, and a very high temperature. The action of the furnace upon the ore divides into three stages.

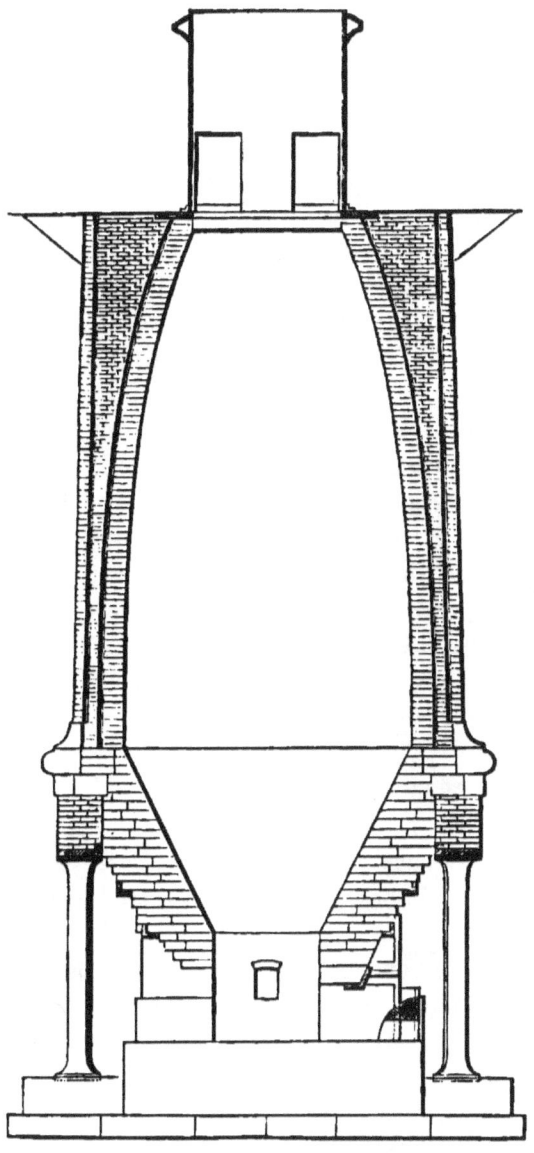

Fig. 54.

I. In the upper portion a **roasting process** takes place, and volatile matters such as water, carbon dioxide, etc., are expelled, and the ore becomes thoroughly calcined. As the combustion goes on below the mass slowly sinks, and is continually replenished from above.

II. In the next stage of the descent the heated ore is **reduced** to the metallic state by the gases of the furnace. In these gases Hydrogen and Marsh gas, CH_4, are present, being produced directly from the coal by heating, but the reduction is chiefly effected by carbonic oxide (CO) produced as follows: In the bottom of the furnace, the blast supplies an excess of air, and the carbon of the fuel is burnt completely into dioxide; but the dioxide in passing upwards is quickly converted by the incandescent carbon to the state of monoxide CO, which coming in contact with iron oxide removes the oxygen and reduces it to the metallic state.

The following shows the composition of one sample of gases drawn from a blast furnace:—

Nitrogen	55	per cent.
Carbon dioxide, CO_2	7	,,
Carbon monoxide, CO	26	,,
Hydrogen	7	,,
Marsh gas, CH_4, etc.	5	,,
	100	

The reducing gases amount in this case to nearly 40 per cent.; the nitrogen is of course introduced in the air of the blast.

III. The iron when first converted into metal is a porous spongy mass, but it soon descends into the hottest portion of the furnace, where it becomes liquefied. In this stage of the operation the iron becomes melted not simply by increase of temperature, but because it enters into combination with **carbon, sulphur, phosphorus,** and **silicon,** producing more easily fusible compounds. The molten metal collects upon the hearth at the bottom of the furnace, with the lighter **slag** floating upon it. The incombustible parts of the coal, together with the limestone, and any silica in the ore, combine and melt into a liquid, which is allowed to flow out over a 'dam-stone' at the bottom of the furnace: this substance when cold has a glassy character, and is known as **slag**. After a sufficient quantity of iron has collected, the furnace is 'tapped,' and the liquid metal permitted to run out into trenches moulded in sand, where it solidifies. The product is termed '**Pig Iron**.'

A great economy of fuel is effected by the use of a *hot* blast of air for the combustion; the heating is done with waste gases from the top of the furnace, which are led down by pipes for the purpose. Further, these waste gases are combustible, and are utilised for raising steam. Tar and ammonia are also obtained as bye products from blast furnace gases.

Purification of Cast Iron.

Pig iron if required for making castings is remelted and run into the moulds; but if malleable or wrought iron is wanted, the iron must be purified by refining and puddling. Pig iron contains besides iron, five or six per cent. of impurities, such as carbon, sulphur, phosphorus, silicon, and manganese; these must be removed to render the iron malleable. The refining process is in reality an oxidising process: the cast iron is remelted and exposed to a blast of air whereby a part of the carbon, silicon, sulphur, and phosphorus are burnt off, and a much purer cast iron obtained.

Wrought Iron.

For the production of wrought iron a charge of cast iron, either crude pig or the refined metal, is brought into a puddling furnace, where it is fused on an open hearth and exposed for some time to the air; as the oxidation of the carbon, etc., proceeds the iron becomes less fusible, and requires frequent stirring and turning over to expose fresh surfaces to the action. Finally, the iron attains a pasty condition and is then withdrawn and hammered, and squeezed between heavy rollers to expel the oxides and slag, next passing on to other pairs of wheels it is rolled into bars or sheets.

Production of Steel.

Good wrought iron contains about 0·3 per cent. of carbon; by an addition of more carbon it is converted into steel. Formerly the change was brought about by packing bars of iron in charcoal dust and keeping them in a furnace of special construction for several days until the steel was formed; but a much more expeditious method is now employed; named after the inventor the **Bessemer** process. For this method the pig iron is melted in cupolas, and run into an egg-shaped vessel termed a **converter**, in which, by means of a blast of air driven through the liquid metal, all the carbon and sulphur are rapidly oxidised and burnt completely out of the metal: at the right instant, when the iron is thus completely decarbonised, by the addition of 'spiegel' (a variety of cast iron) of known composition, the exact quantity of carbon required is put in, and liquid steel produced. This steel is cast into ingots, and rolled into railway bars, etc.

Iron is greyish white metal, capable of taking a high polish. It is extremely tough and tenacious, and very infusible, but at a red heat softens, and can be forged and hammered into any shape. At a white heat iron becomes pasty, and two clean surfaces if hammered together adhere and become perfectly **welded**.

Steel resembles iron in many respects, but is harder, takes a higher polish, and is much more elastic. The hardness and elasticity of steel are modified by the process of **tempering**.

In dry air iron remains bright, but in the presence of moisture becomes rusty, being converted into hydrated oxide, Fe_2O_3, xH_2O: the rusting is accelerated by acid vapours, carbonic acid gas, and salt. At a red heat the surface of iron oxidises in air forming a bluish-black oxide, Fe_3O_4, or ordinary 'smithy scale.'

The compounds of iron are of two kinds, which are very distinct from each other. The lower oxide of iron, **ferrous oxide**, FeO, is basic in character, and forms salts analogous to those of the dyad metal zinc; the higher oxide, **ferric oxide**, Fe_2O_3, is less basic, and forms salts analogous in composition to those of aluminium. The two classes of iron salts are sometimes termed 'proto' and 'per' salts respectively.

The following may be taken as examples of the two classes:—

Ferrous Salts.		Ferric Salts.	
FeO	ferrous oxide.	Fe_2O_3	ferric oxide.
$Fe(OH)_2$	,, hydrate.	$Fe(OH)_3$	,, hydrate.
$FeCl_2$	,, chloride.	$FeCl_3$	,, chloride.
$FeSO_4$	,, sulphate.	$Fe_2(SO_4)_3$	,, sulphate.

Ferrous Oxide, FeO, and Ferrous Hydrate $Fe(OH)_2$, are hardly known in the uncombined state on account of their rapid oxidation in air; the oxide is obtained by gently igniting ferrous oxalate in a tube drawn out to a point and sealing the end when decomposition is complete. The hydrate is thrown down from ferrous salts in solution by alkalies; it is almost white when pure, but rapidly becomes green, and finally red from absorption of oxygen.

An oxide of intermediate composition, Fe_3O_4 or FeO, Fe_2O_3, is found native as magnetic iron ore or 'loadstone.' This oxide is formed as a blackish coating when iron is heated in air or steam.

Ferrous Chloride, $FeCl_2$, may be obtained as a yellowish-white body by acting on iron with gaseous hydrogen chloride at a red heat; iron dissolved in hydrochloric acid yields a green solution, which by concentration deposits greenish crystals, $FeCl_2, 4H_2O$, easily oxidisable in the air.

Ferrous Sulphate, $FeSO_4$. Metallic iron or ferrous sulphide, dissolved in dilute sulphuric acid, yield a pale-green solution, which by crystallisation deposits green crystals of the composition $FeSO_4, 7H_2O$. This salt is known as 'green vitriol' or 'copperas.'

It is manufactured in large quantity from pyrites, FeS_2, which, by exposure to air in heaps, becomes oxidised and converted into an acid solution of ferrous sulphate. This salt is largely used for making writing inks and Prussian blue, and as a mordant in dyeing.

Oxidation of Ferrous Salts.

The conversion of ferrous into ferric salts is effected by numerous agents termed generally 'oxidising' agents, viz. by free oxygen; by many acids containing oxygen, as nitric, chloric, chromic, permanganic, acids, or their salts; by chlorine and bromine in presence of water.

The change is represented thus—
$$2FeSO_4 + H_2SO_4 + [O] = Fe_2(SO_4)_3 + H_2O.$$

Thus to convert ferrous sulphate into ferric sulphate add dilute sulphuric acid to the solution, warm and add gradually a little nitric acid; the green colour of the liquid will change to yellow when the action is complete. The oxidation may also be done by using a solution of potassium chromate or permanganate.

Ferric Oxide, Fe_2O_3. This oxide is found in large deposits as the ore 'hæmatite,' or crystallised as *specular iron*. It is obtained artificially by ignition of ferric hydrate or ferrous sulphate.

Ferric Hydrate, $Fe(OH)_3$, is precipitated as a red gelatinous substance when ferric chloride, or any ferric salt, is treated with alkalies. It dissolves readily in acids, but if dried and ignited loses the combined water, and becomes almost insoluble in acids. A soluble form of ferric hydrate—dialysed iron—is prepared by dialysis of a solution of the chloride.

Ferric Chloride, $FeCl_3$. The anhydrous compound may be obtained by direct combination from the elements. Chlorine gas is passed over gently heated iron, and a brownish-black, volatile, scaly sublimate deposits in a cooler part of the tube. Its vapour density is 78 at $750°C$, corresponding to the molecule $FeCl_3$. This chloride is very deliquescent and unites vigorously with water. Solution of ferric chloride is prepared by dissolving iron in hydrochloric acid and oxidising the ferrous salt with nitric acid [or chlorine]; or iron can be dissolved at once in aqua regia. The solution gives on evaporation a yellowish mass of indefinite composition, which breaks up on further heating into hydrochloric acid and ferric oxide.

Ferric hydrate is soluble to a considerable extent in solution of ferric chloride, but no definite compounds are known.

Reduction of Ferric Salts.

Ferric salts are reduced to the ferrous form by various 'reducing' agents, viz. (1) nascent hydrogen, (2) hydrogen sulphide, (3) sulphurous acid. The change is the converse of the action by which they are formed, being in all cases equivalent to removal of oxygen. The action of hydrogen sulphide is accompanied with separation of sulphur.

$$2FeCl_3 + H_2S = 2FeCl_2 + 2HCl + S.$$

The reduction by nascent hydrogen is effected by dissolving metallic iron or zinc in an acidified solution:—

$$2FeCl_3 + H_2 = 2FeCl_2 + 2HCl.$$

Iron forms two well-defined sulphides, FeS, FeS_2. The **ferrous sulphide** is made by the combination of sulphur with strongly heated iron. It is commonly used for producing sulphuretted hydrogen in the laboratory. The bisulphide or pyrites, FeS_2, is a mineral substance of common occurrence, and frequently associated with copper, arsenic and other metals. Several other sulphides of iron have been artificially made.

Ferrocyanides and Ferricyanides.

Iron with cyanogen forms compounds of great stability, which uniting with other metallic cyanides produce crystallisable salts.

Potassium Ferrocyanide, $K_4FeC_6N_6$, is produced by heating together iron-filings, potassium carbonate, and any nitrogenous organic matter, such as horns, hoofs, etc. It is dissolved out by water, and on evaporation obtained in yellow crystals, sometimes called yellow prussiate of potash. With **ferric salts** it gives a deep-blue precipitate of **Prussian blue**; while with ferrous salts a white or pale-blue precipitate is first formed, rapidly however deepening in colour.

Potassium Ferricyanide, $K_3FeC_6N_6$.

This salt is made by passing chlorine gas through a solution of potassium ferrocyanide.

$$2K_4FeC_6N_6 + Cl_2 = 2K_3FeC_6N_6 + 2KCl.$$

On evaporation red crystals are obtained, sometimes named red prussiate of potash. This salt produces no precipitate with ferric compounds, but with ferrous salts gives a deep-blue precipitate.

Prussian Blue.

If solution of ferric chloride is gradually poured into potassium ferrocyanide, a deep-blue precipitate is obtained containing potassium and iron cyanides:—

$$K_4FeC_6N_6 + FeCl_3 = KFe_2C_6N_6 + 3KCl.$$

Again, if ferrous chloride is similarly poured into potassium ferricyanide a deep-blue precipitate is formed of the same composition:—

$$K_3FeC_6N_6 + FeCl_2 = KFe_2C_6N_6 + 2KCl.$$

This blue substance, when washed, dissolves in pure water, and is named soluble Prussian blue: but treated with solution of **ferric** salt becomes converted into the ordinary insoluble form of **Prussian blue**, Fe_7Cy_{18}, which contains no potassium. The reaction is as follows:—

$$3KFe_2C_6N_6 + FeCl_3 = Fe_7C_{18}N_{18} + 3KCl.$$

If a **ferrous** salt be added to the soluble Prussian blue, a compound named **Turnbull's blue**, Fe_5Cy_{12}, is produced thus:—

$$2KFe_2C_6N_6 + FeCl_2 = Fe_5C_{12}N_{12} + 2KCl.$$

When Prussian blue is heated with potash, part of the iron present separates as ferric hydrate, the rest dissolving to form potassium ferrocyanide:—

$$Fe_7C_{18}N_{18} + 12KOH = 3K_4FeC_6N_6 + 4Fe(OH)_3.$$

Nickel. Atomic weight, 58·6. Specific gravity, 8·9.
Cobalt. Atomic weight, 59·0. Specific gravity, 8·6.

These two metals are associated with each other in nature, and in their chemical and physical properties are singularly alike. There is some little difference in density, but their atomic weights are almost identical. They are usually distinguished by the different colours of their compounds, those of nickel being frequently **green**, while cobalt compounds are commonly **blue** or **red**.

The principal nickel ores are nickel-glance, $NiAsS$, and 'kupfer nickel,' $NiAs$; they are seldom, if ever, free from cobalt.

Nickel forms a series of compounds corresponding to ferrous salts, of which the following will serve as examples:—

Nickel oxide NiO.
Nickel hydrate $Ni(OH)_2$.
Nickel chloride $NiCl_2$.
Nickel sulphate $NiSO_4$.

A few more highly oxidised compounds are known, but they are generally unstable. Nickel peroxide or sesquioxide is Ni_2O_3.

In the arts nickel is used for alloying with copper and zinc to make the hard whitish metal known as 'German silver.' It is also used extensively for electroplating bright iron and other articles, since it takes a high polish and does not readily oxidise.

Cobalt ores resemble those of nickel; the two commonest are cobalt glance, $CoAsS$, and cobalt arsenide, $CoAs_2$. Metallic cobalt is not used in the arts, but an oxide is employed for colouring glass and porcelain. Smalt is a glass coloured blue by cobalt as a silicate.

The affinity of cobalt for oxygen is greater than that of nickel, and the series of cobaltic salts is better represented. It forms two classes of salts corresponding to ferrous and ferric oxides; typified by the following examples:—

Cobalt oxide	CoO.	Cobalt nitrate	CoN_2O_6.
,, hydrate	$Co(OH)_2$.	,, cyanide	CoC_2N_2.
,, chloride	$CoCl_2$.	,, carbonate	$CoCO_3$.

Cobaltic oxide Co_3O_4.

peroxide Co_2O_3.   Cobaltic chloride $CoCl_3$.

Metallic nickel and cobalt are extremely infusible substances and, like iron, are attracted by a magnet. They dissolve in dilute hydrochloric and sulphuric acids with evolution of hydrogen, and the solutions treated with alkali yield the corresponding hydrates, or with sodium carbonate give precipitates of the carbonates.

CHAPTER XXII.

HEXAD METALS. CHROMIUM.

The metals of this group known at present are **Chromium**, atomic weight 52; **Molybdenum**, atomic weight 96; **Tungsten**, atomic weight 184; **Uranium**, atomic weight 240.

Their hexad character is manifested by the formation of oxides of the form R_2O_6 or RO_3, which, like the oxide SO_3, are acid in character, and produce salts by uniting with basic oxides.

K_2SO_4 Potassium sulphate.
K_2CrO_4 ,, chromate.
K_2MoO_4 ,, molybdate.
K_2WO_4 ,, tungstate.
Na_2UO_4 Sodium uranate.

Chromium. Cr. Atomic weight, 52·4. Specific gravity, 7·3.

The chief chromium minerals, which however are not very abundant, are lead chromate, $PbCrO_4$, and chrome ironstone, $FeCr_2O_4$ or FeO, Cr_2O_3.

Besides the acid trioxide CrO_3, chromium forms two lower basic oxides, viz. a sesquioxide Cr_2O_3 generally resembling ferric oxide, and a protoxide, CrO, resembling ferrous oxide. The typical compounds of chromium are accordingly arranged in three classes:—

Chromous Salts.
$Cr(OH)_2$ Chromous hydrate.
$CrCl_2$,, chloride.
$CrSO_4$,, sulphate.

Chromic Salts.
$Cr(OH)_3$ Chromic hydrate.
Cr_2O_3 ,, oxide.
$CrCl_3$,, chloride.
$Cr_2(SO_4)_3$,, sulphate.

Chromates, etc.

CrO_3 Chromium trioxide.
CrF_6 Chromium hexafluoride.
K_2CrO_4 Potassium chromate.
$K_2Cr_2O_7$ Potassium bichromate.

Potassium Chromate, K_2CrO_4.

For the preparation of potassium chromate finely-ground mineral chrome-iron-stone is mixed with potassium carbonate, with some chalk added to keep the mass porous, and heated to redness in a furnace contrived to permit a free access of air. An oxidation takes place, and potassium chromate is produced; this is separated by subsequent washing and crystallisation.

Potassium chromate is a yellow, anhydrous, crystalline salt; soluble in water, making a yellow solution which gives coloured precipitates with many metallic solutions, such as yellow lead chromate, red silver chromate, etc.

Potassium Bichromate. $K_2Cr_2O_7$ or $K_2O, 2CrO_3$.
This salt is produced by adding an acid to potassium chromate. If the yellow solution of chromate be acidified with sulphuric acid, it will change to a reddish-orange colour, when the following reaction occurs:—

$$2K_2CrO_4 + 2H_2SO_4 = K_2Cr_2O_7 + 2KHSO_4 + H_2O.$$

On evaporating the liquid large transparent reddish crystals of the bichromate will be obtained.

The salt dissolves in water, forming a reddish solution, which precipitates metallic chromates. The solution is somewhat acid; if neutralised with potash its colour changes to yellow, when the neutral chromate is formed.

$$K_2Cr_2O_7 + 2KOH = 2K_2CrO_4 + H_2O.$$

Potassium bichromate is much used as an oxidising agent; barium and lead chromates are used as yellow pigments.

Chromium Trioxide. CrO_3. This oxide is easily separated from potassium bichromate. A measured amount of saturated solution of that salt is mixed with a slightly larger quantity of concentrated sulphuric acid, and as the mixture cools the trioxide crystallises out in crimson needles. It is very soluble in water, forming an acid liquid; but true chromic acid, H_2CrO_4, has never been obtained. The trioxide is decomposed by strong heat into chromium sesquioxide and free oxygen.

$$2CrO_3 = Cr_2O_3 + O_3.$$

It is a powerful oxidising substance, and acts energetically upon organic bodies: thus strong alcohol dropped upon the trioxide will take fire.

Reduction of Chromates to Chromic Salts.

Chromates are deprived of oxygen and converted into chromic salts by numerous substances; the action is well shown with sulphurous acid. To obtain a chromic salt it is requisite in all cases to add some acid to combine with the basic chromium oxide into which the reduced chromic acid is converted.

For example:—Make solution of potassium bichromate, acidify it with dilute sulphuric acid, and add gradually a solution of sulphurous acid. As the reduction takes place the reddish liquid becomes a fine green colour; the chemical change is thus expressed:—

$$K_2Cr_2O_7 + H_2SO_4 + 3SO_2 = K_2SO_4 + Cr_2(SO_4)_3 + H_2O.$$

In a similar way chromates are reduced by hydrogen sulphide—with separation of sulphur:—

$$K_2Cr_2O_7 + 4H_2SO_4 + 3H_2S = K_2SO_4 + Cr_2(SO_4)_3 + 7H_2O + S_3.$$

Ferrous salts become ferric; and alcohol (C_2H_6O) becomes changed into aldehyde, C_2H_4O, by the action of chromates; in numerous other cases also bodies are oxidised by chromic acid, which, of course, suffers reduction to a chromic salt.

Chromic Salts.

Chromium Hydrate. $Cr(OH)_3$. The green solution obtained in the manner just described by the action of sulphurous acid upon a chromate and containing chromic sulphate, if treated with alkalies, gives a precipitate of chromic hydrate or hydrated sesquioxide of chromium, as a greenish gelatinous body. It dissolves in acids, forming chromic salts, and when heated leaves anhydrous chromic oxide, Cr_2O_3.

Chromium Sesquioxide, Cr_2O_3, thus obtained is a green amorphous substance, not poisonous, and being very permanent is frequently used as a pigment.

Chrome Alum. $KCr(SO_4)_2, 12H_2O$. This alum is most easily obtained by reducing potassium bichromate; e. g. with sulphurous acid.

$$K_2Cr_2O_7 + 4H_2SO_4 - [O_3] = [K_2SO_4, Cr_2(SO_4)_3] + 4H_2O.$$

If the green solution be evaporated and allowed to stand some days, a gradual formation of reddish violet-tinted crystals of chrome alum takes place. They dissolve in cold water, forming a reddish-violet solution, which if heated becomes again of a green colour.

Chromous Salts. The whole of these compounds are difficult to keep on account of their extreme liability to oxidation. **Chromous chloride,** $CrCl_2$, which may be taken as an example of the series, is left when hydrogen is passed over heated chromic chloride:—

$$2CrCl_3 + H_2 = 2CrCl_2 + 2HCl.$$

They are like ferrous salts in their general characters, reactions, and typical compounds.

Oxidation of Chromic Salts to Chromates. Chromic oxide, Cr_2O_3, and other compounds, if fused with potassium nitrate, are oxidised, yielding potassium chromate, and a similar change is effected by boiling with potassium chlorate in strong nitric acid.

P

CHAPTER XXIII.

HEPTAD METALS.

Manganese. Mn. Atomic weight, 55. Specific gravity, 8·0.

THE only metal of this class is manganese, which is regarded as a heptad on account of the formation of permanganates, $KMnO_4$ or $K_2Mn_2O_8$; these, like perchlorates and periodates, appear to contain a heptad oxide, viz. Mn_2O_7:—

$$K_2O, Mn_2O_7 = K_2Mn_2O_8.$$

In respect of its atomic weight and general characters manganese is closely allied to iron, and commonly is found associated with iron in nature. The protoxide of manganese, and the salts derived from it, are scarcely distinguishable from the corresponding ferrous salts, except in colour; crystallised manganous salts being mostly pink, while the ferrous salts are green.

The following are some of the most important compounds of this metal:—

Manganese protoxide	MnO.
Manganese dioxide	MnO_2.
Manganese sesquioxide	Mn_2O_3.
Red manganese oxide	Mn_3O_4.

Manganous sulphate	$MnSO_4$.	Manganous chloride	$MnCl_2$.
Manganous sulphide	MnS.	Manganous carbonate	$MnCO_3$.
Potassium manganate	K_2MnO_4.		
Potassium permanganate	$KMnO_4$.		

The most common and abundant ore of manganese is the dioxide, MnO_2, a dark grey or black mineral, known as **Pyrolusite**; the lower oxides Mn_2O_3, or **Braunite**;

and Mn_3O_4, or **Hausmannite**, being less frequently found.

But manganese is also found in small quantities in most soils, and is usually present in ores of iron; and consequently is generally to be detected in specimens of manufactured iron. A variety of cast iron rich in manganese and carbon, known as ferro-manganese, is added to the contents of the crucible in the manufacture of Bessemer steel.

Metallic manganese can be obtained by reducing the oxide with carbon; it is a grayish white metal—in appearance resembling cast iron. In the air it oxidises more readily even than iron, and is usually preserved in naphtha or in vacuous tubes. Manganese requires an extremely high temperature to melt it. The metal is soluble in dilute acids with evolution of hydrogen, and producing manganous salts.

Manganous Chloride. $MnCl_2$. Black oxide of manganese is digested with hydrochloric acid until the evolution of chlorine ceases, when a solution of manganous chloride is obtained.

$$MnO_2 + 4HCl = MnCl_2 + Cl_2 + 2H_2O.$$

Iron is commonly present in the mineral oxide, and dissolves also in the acid forming ferric chloride: the addition of a little powdered manganous carbonate will precipitate all the iron, leaving a pure solution of manganese chloride.

$$2FeCl_3 + 3H_2O + 3MnCO_3 = 2Fe(OH)_3 + 3MnCl_2 + 3CO_2.$$

On evaporating the liquid pinkish crystals of manganous chloride will be obtained, $MnCl_2, 4H_2O$.

Manganous Sulphate. $MnSO_4$. This salt is obtained in a similar way: black oxide of manganous is heated with strong sulphuric acid (oxygen being given off), and the manganous sulphate separated by solution and crystallisation. Iron, if present, can be removed by precipitation with manganous carbonate.

Solutions of manganous sulphate or chloride are decomposed by alkalies the hydrate being precipitated.

$$MnSO_4 + 2KOH = Mn(OH)_2 + K_2SO_4.$$

Manganous hydrate is nearly white, but rapidly turns brown from absorption of atmospheric oxygen.

When ammonium sulphide and ammonium hydrate are added to solutions of manganous salts a flesh-coloured precipitate of hydrated manganous sulphide is produced.

Potassium Manganate. K_2MnO_4.

When a manganese compound is heated with an alkali in presence of oxygen, a green mass of *manganate* is formed; the action is analogous to the formation of chromates.

$$MnO_2 + [O] + 2KOH = K_2MnO_4 + H_2O.$$

The potassium salt is usually prepared as follows:— Manganese peroxide and potassium chlorate are added to a strong solution of potash, and the mixture heated until a dry mass is obtained. This mass is further heated to a low red heat in an iron vessel until it becomes dark green in colour: the dull green substance is potassium manganate. Cold water dissolves it, forming a green solution, which is very unstable, since it decomposes on simply heating, or by the addition of acids, and a purple red solution of **permanganate** is obtained. The change of colour is very striking, and from the way in which it takes place, the manganate has received the name of Chamæleon Mineral.

Potassium Permanganate, $KMnO_4$.

The green solution of manganate prepared in the

manner already described, is converted into permanganate by boiling, or by the action of an acid:—

$$3K_2MnO_4 + 3H_2O = 2KMnO_4 + MnO_2 + 4HKO + H_2O.$$

The potassium manganate with water forms potassium permanganate, manganese dioxide, potash and water, and the manganese oxide precipitates as a dark brown powder. With an acid no precipitation takes place:—

$$5K_2MnO_4 + 4H_2SO_4 = 4KMnO_4 + MnSO_4 + 3K_2SO_4 + 4H_2O.$$

Potassium permanganate is formed, with manganous sulphate, and potassium sulphate. The pure salt is obtained by evaporation and crystallisation; it forms needle-like crystals of bronzy lustre, and is in every way a more stable compound than the manganate. Crude solutions of manganate and permanganate are used for disinfecting purposes (Condy's fluid). Since permanganates contain much oxygen, they are useful as oxidising agents; and accordingly are de-oxidised by reducing agents, such as those capable of reducing chromates. For example, a solution of permanganate mixed with a ferrous salt loses its red colour, the reaction being as follows:—

$$2KMnO_4 + 3H_2SO_4 = 2MnSO_4 + K_2SO_4 + [O_5] + 3H_2O.$$

Two molecules of permanganate with sulphuric acid become changed into manganous sulphate, potassium sulphate, and five atoms of available oxygen, which is, of course, taken up by the ferrous salt in its conversion to a ferric salt:—

$$2FeSO_4 + H_2SO_4 + [O] = Fe_2(SO_4)_3 + H_2O.$$

Cuprous salts, mercurous salts, and stannous salts; sulphurous acid, hydrogen sulphide, potassium nitrite, and potassium iodide in a similar manner reduce and decolorise a solution of the permanganate.

CHAPTER XXIV.

METALS OF THE COPPER GROUP.

Copper, Silver, Mercury.

The three metals copper, silver, and mercury, form a fairly definite natural group, and yield derivatives of similar characters and composition. In respect of their lower oxides they resemble the monad metals, for example, the oxides Cu_2O, Ag_2O, Hg_2O, are comparable with K_2O; but the higher oxides and chlorides are dyad in type, and CuO, HgO (and possibly AgO) are comparable with CaO, and the chlorides $CuCl_2$, $HgCl_2$ with dyad chlorides. The metals themselves are easily obtained by reduction of their salts, and are permanent in air. Copper and mercury form salts and other compounds of two degrees of oxidation, but in the case of silver the higher series is represented by very few compounds.

The lower compounds of these metals with chlorine, bromine, and iodine are peculiarly sensitive to the action of light; a character which in the case of certain silver salts is utilised for the production of photographic pictures.

Copper. Cu. Atomic weight, 63. Specific gravity, 8·9.

The most important ores of copper are the sulphide, or copper pyrites CuS, containing variable quantities of iron; the basic carbonate or malachite, $Cu_2(OH)_2CO_3$; and the red oxide Cu_2O. The metal is occasionally found native. Copper is widely distributed in nature; frequently small quantities are found in the ores of other metals, and traces of this element are found in many plants, seeds, and in some animal tissues.

The smelting of copper ore in the form of carbonate or oxide is performed by reduction with carbon; but for the smelting of pyrites or sulphides a more complex process is necessary. The ore is first roasted or calcined in air, during which much sulphur and arsenic burn off, as sulphur dioxide, and arsenic trioxide, and the iron oxidises to ferric oxide. The copper sulphide loses also some sulphur, and cuprous sulphide is left:—

$$2CuS + O_2 = Cu_2S + SO_2.$$

By fusion the impure cuprous sulphide is separated, and is termed technically **coarse** metal. A repetition of the roasting with the addition of slag from other operations, produces nearly pure cuprous sulphide, termed fine metal, which is almost free from iron. The fine metal is fit for the final stage of reduction, which is carried out as follows: the cuprous sulphide is heated in a furnace to which air is admitted only in sufficient quantity to oxidise a portion of the mass; the heat is then increased until the oxide and sulphide react with each other, and the metal is liberated. The chemical change is expressed by the equation:—

$$Cu_2S + 2CuO = 4Cu + SO_2.$$

Cuprous sulphide and cupric oxide yield metallic copper and sulphur dioxide. The copper thus obtained is known as 'blister copper,' and is rendered somewhat brittle by the presence of some oxide. After being remelted and refined by stirring with a pole of green wood, a tough and fibrous metal is obtained.

Copper is a bright red-coloured metal, rather soft,

but very tough and malleable, so that it admits of being hammered and worked into vessels of various forms. It fuses at a very high temperature, and can be alloyed in the molten state with many other metals. With half its weight of zinc it forms **brass**, with different proportions of tin the varieties of bronze, bell-metal, and gun-metal are produced. Our English bronze coinage contains 95 per cent. of copper, with 4 per cent. of tin, and 1 per cent. of zinc. Alloyed with aluminium the useful golden-coloured alloys known as 'aluminium bronze' are produced.

Copper is an excellent conductor of electricity, and is consequently largely used in connection with telegraphy, electric lighting, and for similar purposes.

Polished copper does not oxidise to any extent in dry air at ordinary temperatures, but becomes tarnished with a superficial coating only of oxide; when heated to redness, however, the oxidation of the metal is rapid.

Copper dissolves readily in nitric acid producing a nitrate, and in aqua regia producing a chloride; but with hydrochloric acid the metal is acted on very slowly. In a fine state of division it dissolves in the boiling acid with evolution of hydrogen.

Copper compounds are divided into two classes—**cuprous** and **cupric**, of which the following are types:—

Cuprous compounds.	Cupric compounds.
Cuprous oxide Cu_2O.	Cupric oxide CuO.
,, chloride Cu_2Cl_2.	,, chloride $CuCl_2$.
,, iodide Cu_2I_2.	,, sulphate $CuSO_4$.
,, sulphide Cu_2S.	,, nitrate CuN_2O_6.

Cuprous Chloride. Cu_2Cl_2.

This compound is easily obtained by digesting together copper oxide, metallic copper, and excess of strong hydrochloric acid; a dark brown liquid containing cuprous chloride and hydrochloric acid is produced.

$$CuO + Cu + 2HCl = Cu_2Cl_2 + H_2O.$$

Upon diluting the strong solution with water, cuprous chloride separates out as a white powder: this chloride is not soluble in water, but dissolves in strong solutions of hydrogen chloride, or the chlorides of sodium, potassium, and ammonium, forming colourless liquids. These solutions, which contain a double chloride ($KCuCl_2$?), are decomposed by excess of water, and the cuprous compound precipitated. An important property of the solutions is their ability to absorb **carbonic oxide** gas, for which purpose they are used in the analysis of gaseous mixtures. The cuprous solutions absorb oxygen when exposed to air, by chlorine water they are converted into green cupric salts.

Cuprous Oxide. Cu_2O. This oxide, frequently called sub-oxide or red oxide of copper, is obtained by reducing cupric solutions with organic bodies. A solution of copper sulphate, for example, is mixed with **grape sugar**, and caustic alkali (soda or potash) added in excess; a clear solution of a deep blue colour is obtained, which on warming deposits cuprous oxide varying in tint from yellow to orange and bright red. By the use of a standard solution of copper this reaction may be employed in the quantitative estimation of sugars. Red oxide of copper is used for giving a ruby red colour to stained glass.

Cupric Oxide. CuO. This is the black copper oxide, which may be obtained by igniting copper nitrate, or by heating the metal to redness in air. It is a stable compound, but when heated to redness in hydrogen, or in carbonic oxide, becomes reduced to the metallic state.

Organic Analysis.

Cupric oxide is employed in **organic analysis**, as when an organic substance is mixed with the oxide, and heated to redness, the carbon is burnt to carbon dioxide, and the hydrogen to water, metallic copper being left. Thus, let a weighed amount of sugar be mixed with excess of copper oxide, and the mixture brought into a hard glass combustion tube. This is connected with a ∪-tube containing dry calcium chloride for absorbing the water formed, and also a bulb apparatus containing solution of potash for taking up the carbon dioxide. The combustion-tube, when thus prepared, is placed in a suitable furnace and heated to redness, and by weighing accurately the two pieces of absorption apparatus before and after the combustion, the amounts of water and carbon dioxide generated are ascertained. From these data the carbon and hydrogen in sugar can be calculated, and the oxygen being found by difference, the percentage composition and also the formula of the sugar are found.

An actual example will make this clear. ·2 grams of substance when burnt with cupric oxide produced ·294 grams of CO_2 and ·120 grams of H_2O. From these data we are able to find out what is the per centage composition of the substance (which contains only carbon, hydrogen, and oxygen) and to calculate its formula.

Since 44 parts of CO_2 contain 12 of C., the amount of carbon in the sugar is ·08 grams or 40 per cent., and the hydrogen present being one-ninth of the water formed, is ·013 grammes or 6·66 per cent. We get therefore:—

        Carbon   = 40·00 per cent.
        Hydrogen = 6·66   „
        Oxygen   = 53·34   „   by difference,

and dividing each percentage by the atomic weight of the element we get a formula $C_{3\cdot3} H_{6\cdot6} O_{3\cdot3}$, or in whole numbers CH_2O. The formula of the substance burnt may be any multiple of these numbers, as $C_2H_4O_2$: $C_6H_{12}O_6$, etc.

Cupric Hydrate. $Cu(OH)_2$. When fixed alkalies, potash, or soda are added to solutions of cupric salts, such as the sulphate or chloride, a pale blue gelatinous precipitate of cupric hydrate is thrown down; which, if heated to the boiling temperature, is converted into the black oxide.

Cupric Chloride. $CuCl_2$. By direct union of copper and chlorine a brown chloride is formed, which dissolves in water, producing a pale blue solution, and from the solution a bluish green hydrated salt is obtained by evaporation having the composition, $CuCl_2, 2H_2O$. Cupric oxide or the carbonate may also be dissolved in hydrochloric acid to form the chloride.

Cupric Sulphate. $CuSO_4$. This is the common substance known by the name of 'blue vitriol.' It is prepared on the large scale by roasting copper pyrites at a low temperature, when by absorbing oxygen it is converted into the sulphate.

$$CuS + O_4 = CuSO_4.$$

The crude sulphate is dissolved in water, and obtained by crystallisation in large blue crystals containing $CuSO_4, 5H_2O$.

The sulphate is also formed by acting on copper with strong sulphuric acid:—

$$Cu + 2H_2SO_4 = CuSO_4 + SO_2 + 2H_2O,$$

or otherwise by the solution of cupric oxide in sulphuric acid.

Upon heating the blue crystals until they lose all their water anhydrous copper sulphate is left as a white powder; it recombines eagerly with water, becoming hot if wetted and changing again to a blue colour.

Cupric Nitrate. CuN_2O_6. This salt is formed when copper dissolves in nitric acid, as in the preparation of nitrogen dioxide. On evaporation of the blue solution the nitrate is obtained in deliquescent blue crystals. It is very soluble in water, and when heated breaks up into black cupric oxide, free oxygen, and nitrogen oxides.

Electrotyping with Copper. When a current of electricity is passed through a solution of copper sulphate, a decomposition takes place, and copper is deposited upon the negative pole, while sulphuric acid is liberated at the positive pole, and oxygen is evolved; this however only describes the effect when platinum poles are used. But if a sheet of copper is used as the positive pole sulphuric acid is not liberated, since it attacks and dissolves the copper pole, and as a consequence the strength of the solution keeps constant. So long as the current continues a deposition of copper goes on at one pole and a solution of copper at the other; the copper being deposited in a coherent film over the surface of any object attached to the negative pole. Thus an identical reproduction or cast of the object is produced, and by this method copies of wood engravings, letterpress, etc., are obtained with perfect accuracy.

Silver. Ag. (Argentum). Atomic weight, 108. Specific gravity, 10·5.

Silver is sometimes found native, but more usually in association with sulphur, and antimony, arsenic, or lead sulphides; the chloride is occasionally met with as a mineral.

The extraction of silver from the sulphide is effected by roasting with salt, so as to convert the metal into chloride; the sulphur becomes oxidised to sulphate, which combines to form sodium sulphate:—

$$Ag_2S + 2NaCl + 2O_2 = 2AgCl + Na_2SO_4.$$

Any copper or iron in the ore becomes converted also in part to chloride, or in part to sulphate.

The silver is next separated by the process of **amalgamation** with mercury. The roasted mass of chlorides, etc., is put into revolving tubs with water and scraps of iron, and kept in motion. The iron reduces the chloride to metallic silver, and some mercury being added to the contents of the cask, the reduced silver is dissolved, forming an amalgam with the mercury: by a subsequent distillation the two are separated, the volatile mercury being driven off, leaving a residue of silver, and usually also some copper.

Most lead ores contain silver, which is extracted along with the lead. The method of separating the silver from lead is described in the chapter upon the latter.

Silver is a beautiful white metal, capable of taking a high polish; it does not oxidise in air, but after a time becomes tarnished and blackened from the formation of sulphide. The melting-point of silver is 960° C., and it boils at the temperature of about 1600° C.; it can be distilled in the oxy-hydrogen blowpipe flame.

Silver is a malleable, ductile metal, and silver wire is the best-known conductor of electricity.

When used for coinage, silver is alloyed with copper; thus in English coinage 7·5 per cent. of copper is used, in French 10 per cent.

Silver is made into cups and tankards, spoons and forks, by hammering and stamping; but for many purposes objects are coated with a film only of the metal, the *plating*, as it is termed, being soldered on.

Electroplating. Metal articles are plated with silver by the electric current. A solution of silver cyanide in potassium cyanide is used; the object to be silvered is attached to the negative pole, and a plate of silver used on the positive pole, so as to maintain the strength of the solution. The quantity of silver thrown down at one pole during the passage of the current is equal to the quantity dissolved at the other.

Silvering on Glass. Silver salts may be reduced to the metallic state by numerous organic substances, such, for example, as tartrates, milk sugar, aldehyde, etc., and under suitable conditions the silver separates as a bright metallic coating. Thus, if a few grains of calcium tartrate are placed in a test tube with a little weak ammonia, and a crystal of silver nitrate added, upon gently warming the tube a brilliant film of metallic silver will form on the glass.

Mirrors are now frequently made by silvering instead of being coated with mercury-tin alloy. Solution of silver nitrate, with a little ammonia, is made alkaline with pure potash and grape sugar added; the silver deposits upon the glass without heating, and after washing and drying is very coherent. It is usually protected afterwards by a coating of varnish applied to the back.

Silver dissolves readily in nitric acid, but on account of the insolubility of its chloride, will not dissolve in hydrochloric acid or aqua regia. Strong sulphuric acid converts it into sulphate.

The following are some typical silver compounds:—

Ag_2O Silver oxide.
$AgCl$ „ chloride.
$AgNO_3$ „ nitrate.
Ag_2SO_4 „ sulphate

A peroxide of silver [Ag_2O_2 or AgO] has been obtained; it is the only silver compound known corresponding to cupric oxide.

Silver is represented in its compounds as a monad element like potassium, but their vapour densities not being known, the true molecular weight of the salts of silver is not yet ascertained with certainty. An experiment recently performed gives 160·8 as the density of vapour of silver chloride, the theoretical value being 143·5 for AgCl.

Silver Nitrate. $AgNO_3$. Nitric acid dissolves silver readily, nitric oxide is given off, and the nitrate formed. The salt is soluble in water, and is obtained by evaporation in colourless crystals, which are anhydrous. It fuses at a low heat, and when cast into sticks is known as 'lunar caustic,' being used as a caustic for surgical purposes. A solution of silver nitrate is used in the laboratory as a reagent, since many acid radicles by forming insoluble silver salts produce characteristic precipitates. For example:—

$$2 AgNO_3 + K_2CrO_4 = Ag_2CrO_4 + 2 KNO_3.$$

The crimson silver chromate is formed by mixing a soluble chromate with solution of silver nitrate.

Silver Chloride. AgCl. By adding to solution of silver nitrate any soluble chloride, the silver is precipitated in combination with chlorine.

The chloride of silver is a curdy, white substance, insoluble in nitric acid, but readily dissolved by ammonia. The bromide and iodide, which may be obtained in a similar manner, are yellowish in colour, and the latter compound is not dissolved by ammonia. All three are soluble in solution of sodium hyposulphite,—a fact of much importance in the production of photographic pictures.

The compounds of the halogen elements with silver are decomposed by exposure to light to a remarkable extent, and this property is the basis of many photographic processes. Chloride of silver alone, if exposed to light, is seen to change colour to violet and black, but in the presence of organic bodies the change takes place far more rapidly.

A glass plate coated with a film of gelatine containing particles of silver bromide, if exposed in a camera for a small fraction of a second to the image of a landscape focussed upon the plate, is chemically affected by the light falling on it, and by means of **reducing agents**, such as ferrous oxalate or pyrogallic acid, a picture can be **developed** in the film, varying in light and shade just as the image allowed for an instant to fall on it varied in its different parts.

Mercury. Hg. Hydrargyrum. Atomic weight, 200. Molecular weight, 200. Specific gravity, 13·6.

The metal mercury, like the others of the group, is found in the free state, but it appears not to be as widely distributed in nature as copper and silver. The principal ore is cinnabar (HgS), but mercurous chloride (Hg_2Cl_2), as well as the iodide (Hg_2I_2), and selenide ($HgSe$) are sometimes found. Metallic mercury is readily obtained from the sulphide: a mixture of cinnabar and lime is heated, and the mercury distilling off is collected in receivers.

$$HgS + CaO = Hg + O + CaS.$$

Calcium sulphide is thus obtained, but becomes in part oxidised to sulphate, $CaSO_4$. Sometimes the separation of mercury is accomplished simply by cautious oxidation of the ore:—

$$Hg + SO_3 = Hg + SO_3.$$

Mercury is a bright white metal, liquid at ordinary temperatures. It solidifies at $-39°$ C., and can be frozen by a mixture of solid carbon dioxide and ether, or by the evaporation of liquefied gases, such as nitrous oxide, ethylene, etc. At $385°$ C. it boils, and can be purified from traces of other metals by the process of distillation. The density of mercury vapour is 100 times that of hydrogen, so that the molecular weight of the element is 200; for 200 parts by weight of mercury vapour occupy the space of 2 parts by weight of hydrogen, $-Hg : H_2$. The atomic weight of mercury is 200 also, and thus mercury resembles the dyad metals zinc and cadmium in having both molecule and atom of the same mass.

Mercury is a good conductor of electricity, and alloys with other metals producing **amalgams** of a liquid or pasty character, according to the proportion of mercury in the mixture. It combines energetically with sodium and potassium, with evolution of heat and light. The best solvent for mercury is nitric acid; it dissolves also in aqua regia, forming perchloride, but hydrochloric acid is almost without action on it. Strong sulphuric acid converts the metal into mercurous sulphate, sulphur dioxide being liberated. Mercury forms two well-defined series of compounds—**mercurous** and **mercuric**, of which the following may serve as examples:—

Mercurous compounds.			**Mercuric compounds.**		
Hg_2O	Mercurous	oxide.	HgO	Mercuric	oxide.
Hg_2Cl_2	,,	chloride.	$HgCl_2$	,,	chloride.
$Hg_2N_2O_6$	,,	nitrate.	HgN_2O_6	,,	nitrate.
Hg_2SO_4	,,	sulphate.	$HgSO_4$	,,	sulphate.
			HgS	,,	sulphide.

Mercurous Oxide. Hg_2O. Mercurous oxide is produced by the action of potash or other alkali upon mercurous salts. If added to a solution of mercurous nitrate, potash produces a black precipitate of this oxide. It breaks up readily into metal and mercuric oxide.

Mercuric Oxide. Hg O. Red oxide of mercury or **Red Precipitate.** This is the oxide formed by the prolonged heating of mercury in air: thus made it is a dull red crystalline powder. If mercury nitrate (alone or mixed with mercury) be heated, the oxide is left as bright red crystalline scales, but when a soluble mercuric salt, for example mercuric chloride, is mixed with alkali (soda or potash), the oxide is thrown down as a **yellow amorphous** powder. Both red and yellow oxides are used in medicine. All the forms give when heated oxygen and metallic mercury.

Mercurous Chloride. Hg_2Cl_2. Calomel.

The lower chloride of mercury, known as calomel, can be formed from the elements, or is produced as a white precipitate by mixing mercurous nitrate with hydrochloric acid or a soluble chloride, such as sodium chloride:—

$$Hg_2N_2O_6 + 2HCl = Hg_2Cl_2 + 2HNO_3.$$

It is commonly made by subliming a mixture of mercuric chloride and metallic mercury.

$$HgCl_2 + Hg = Hg_2Cl_2.$$

Or otherwise a mixture of mercurous sulphate and common salt is gently heated until the calomel sublimes.

$$Hg_2SO_4 + 2NaCl = Na_2SO_4 + Hg_2Cl_2.$$

Since the calomel thus made may contain traces of the soluble poisonous chloride ($HgCl_2$), it must be well washed to dissolve out that substance.

Calomel is a white powder, becoming buff or brownish in colour when exposed to light. It is not soluble in water, and when heated sublimes without fusion; potash converts it into black oxide, and it also blackens with ammonia. Aqua regia dissolves calomel, converting it into the higher chloride, $HgCl_2$, but hydrochloric acid decomposes it into mercury and mercuric chloride. Calomel is used in small doses as a purgative medicine.

Mercuric Chloride. $HgCl_2$. Corrosive Sublimate.

Mercuric chloride may be obtained by allowing mercury to combine with excess of chlorine, but it is commonly made by sublimation. A mixture of **mercuric sulphate** and **salt** is heated in a glass flask placed in a sand bath; the action is as follows:—

$$HgSO_4 + 2NaCl = HgCl_2 + Na_2SO_4,$$

the sodium sulphate formed in the action is left, and the mercury compound volatilised into the upper part of the vessel.

Mercuric chloride is a colourless crystalline salt, soluble in water, and also in alcohol. It is a violent and dangerous **poison**; and indeed this is a character of many mercury compounds. A solution of mercury chloride mixed with stannous chloride yields first a white precipitate of calomel, and finally a grey powder of metallic mercury.

$$2HgCl_2 + SnCl_2 = Hg_2Cl_2 + SnCl_4.$$
$$HgCl_2 + SnCl_2 = Hg + SnCl_4.$$

Solution of mercurous chloride has great antiseptic properties, as it is fatal to the organisms causing putrescence, and it is much used for the preservation of anatomical specimens, and for preserving skins and furs.

Solid mercuric chloride melts at $265°$ C., and boils at $295°$ C.; its vapour density is 135.5, and the molecule therefore weighs 271 and is represented by $HgCl_2$.

White Precipitate. $HgClNH_2$. When ammonia is added to solution of corrosive sublimate, a white precipitate is obtained, in which half the chlorine originally present is replaced by the radicle NH_2. It is therefore called mercuric chlor-amide, and is formed according to the following equation:—

$$HgCl_2 + 2NH_3 = HgCl, NH_2 + NH_4Cl.$$

The black substance produced when solution of ammonia is poured upon calomel is similar in character, and is mercurous chlor-amide, Hg_2Cl, NH_2.

Molecular Weight of Calomel.

Much difference of opinion has existed as to the molecular formula of calomel; whether it should be written $235 \cdot 5 = HgCl$, or as $471 = Hg_2Cl_2$. By numerous experiments it has been shown that the vapour-density is about 118, which corresponds to the molecule $HgCl$; but, on the other hand, it has been shown that calomel vapour amalgamates gold leaf and apparently dissociates by heating into mercuric chloride and mercury. If the dissociation were complete we should get:—

$$\begin{array}{ccc} Hg_2Cl_2 & = & Hg & + & HgCl_2. \\ [\text{2 vols.}]? & & \text{2 vols.} & & \text{2 vols.} \end{array}$$

The dissociation, however, is certainly limited in amount, and not complete as represented by the equation: and the vapour-density does not vary in such a way as to support the hypothesis of gradual dissociation. We may therefore suppose that the molecule of calomel vapour is really represented by $HgCl$, which may be more or less dissociated without change of density—

$$\begin{array}{ccc} 2HgCl & = & Hg & + & HgCl_2. \\ 2 \times \text{2 vols.} & & \text{2 vols.} & & \text{2 vols.} \end{array}$$

Experiments on the density of the vapour of a mixture of calomel and corrosive sublimate give for calomel a value about 118, and so support this view. In the case of such mixtures, too, no free mercury could be detected in the vapour when it came into contact with a gilt surface.

Mercury Salts.

Mercurous Nitrate, $Hg_2N_2O_6$.

In contact with cold dilute nitric acid, metallic mercury is converted into mercurous nitrate; but if the metal be heated with strong nitric acid, mercuric nitrate is formed. There are several basic mercurous nitrates known besides the normal salt which crystallises with two molecules of water, $-Hg_2N_2O_6, 2H_2O$. Thus the normal mercurous nitrate decomposes with water, forming a yellow basic nitrate, $Hg_4H_2N_2O_8$:—

$$2Hg_2N_2O_6 + 2H_2O = Hg_4H_2N_2O_8 + 2HNO_3.$$

Another subnitrate in large transparent crystals is formed by digesting an excess of mercury with dilute nitric acid; its composition is:— $Hg_8H_2N_6O_{20}$ or $4Hg_2O, 3N_2O_5, OH_2$.

Solutions of mercurous nitrates give, with hydrochloric acid, a white precipitate of calomel or mercurous chloride; with alkalies they give the black mercurous oxide.

Mercuric Nitrate. HgN_2O_6. This mercuric salt is formed by long continued heating of the solution made by dissolving mercury in excess of nitric acid, but may be conveniently obtained by dissolving mercuric oxide in nitric acid:— .

$$HgO + 2HNO_3 = HgN_2O_6 + H_2O.$$

It crystallises from a strong acid solution in transparent crystals, which are very deliquescent. When much diluted with water the salt is decomposed, and a basic salt left, which by long washing with hot water finally loses all its nitric acid, leaving only mercuric oxide.

Vermilion, Mercuric Sulphide. HgS. The mineral sulphide of mercury is known as cinnabar; it is of a dull red colour, becoming brighter when rubbed. If mercury and sulphur are triturated together a **black**

amorphous sulphide [Aethiops' mineral] is obtained; but when cautiously sublimed, the compound is converted into the **red crystalline** modification known as Vermilion, which is identical in composition with the other variety. Vermilion is also prepared in the wet way: by acting upon the black sulphide for some time with potash containing dissolved sulphur [in the form of potassium pentasulphide], it slowly becomes converted into a fine bright scarlet powder. Vermilion is used as a pigment in painting and for colouring sealing-wax.

CHAPTER XXV.

TETRAD METALS. TIN AND LEAD.

The metals Tin and Lead belong to the group of tetrad elements both forming dioxides. Tin too forms a definite tetrachloride, $SnCl_4$, and the formation of the organic compounds tin tetrethyl, $Sn(C_2H_5)_4$, and lead tetrethyl, $Pb(C_2H_5)_4$, in which the ethyl group, C_2H_5, is a monovalent radical, is also indicative of tetrad character.

Both metals can be readily reduced from their oxides, and are permanent in air at ordinary temperatures; they are, however, fusible at low temperatures, and when fused rapidly become converted into oxides.

Tin. Sn. Stannum. Atomic weight, 118. Specific gravity, 7·3.

The only ore of tin of practical importance is Tin Stone, SnO_2, which is found in a few localities, chiefly in Cornwall, Malacca, and Australia. To extract the tin the ore is crushed to powder and washed in running water; the heavy tin stone is left, while the lighter impurities are carried away. It is next roasted, in order to burn away the arsenical iron pyrites present; arsenic and sulphur are thus volatilised as oxides, ferric oxide being left. A second washing removes most of the iron oxide, and fairly pure tin dioxide is obtained: this is reduced to metal by heating with powdered charcoal.

Tin is a bright white metal, with good metallic lustre; it melts at 235°C., and volatilises only at a high temperature. The metal is obscurely crystalline, and crackles when bent; it becomes brittle when hot, and at about 200° may be readily powdered.

Tin is much used for coating copper vessels used in cooking, and for covering sheet iron, which is then known as 'tin plate.' If a clean surface of either copper or iron is dipped into molten tin, the metal adheres very strongly. An amalgam of tin foil and mercury was commonly used for coating glass mirrors, but the 'silvering' is now more generally done with silver reduced from solution by means of organic substances.

With **lead**, tin produces the fusible alloys known as **pewter** and **solder**; mixed with **copper** in varying proportions it makes **gun metal**, **bell metal**, and **bronze**.

Metallic tin is soluble in hydrochloric acid, forming stannous chloride, and in aqua regia forming stannic chloride; but nitric acid converts it into a white powder consisting of the hydrated dioxide.

The compounds of tin are divided into two classes, of which the following substances are typical:—

Stannous compounds.			**Stannic compounds.**		
SnO	Stannous	oxide.	SnO_2	Stannic	oxide.
$Sn(OH)_2$	,,	hydrate.	$Sn(OH)_4$	,,	acid.
$SnCl_2$	,,	chloride.	$SnCl_4$	,,	chloride.

Stannous Chloride, $SnCl_2$.

Metallic tin dissolves in hot hydrochloric acid, hydrogen being liberated, and the solution on evaporation deposits crystals of hydrated stannous chloride, $SnCl_2, 2H_2O$. The water can be expelled by heating the crystals in a closed vessel out of contact of air; but if the heating is done in an open vessel, oxygen is absorbed and an oxychloride produced. The anhydrous chloride is formed by heating a mixture of tin filings with mercuric chloride. The mercury is first distilled off, and afterwards the stannous chloride at a red heat can be volatilised. The vapour density of stannous chloride at high temperatures indicates a molecular weight of $189 = SnCl_2$.

Stannous chloride absorbs both oxygen and chlorine eagerly, and is often employed as a reducing agent. It completely reduces salts of mercury, silver, and gold to the metallic state, and chromates, manganates, etc., are brought to a lower stage of oxidation. Ferric and cupric compounds are also reduced to the ferrous and cuprous states. Crystallised double salts are formed by the union of stannous chloride with the chlorides of potassium, ammonium, etc.; such, for example, as K_2SnCl_4.

Stannous Oxide. SnO. This oxide is left on heating stannous oxalate in a tube drawn out to a point, which allows gas to escape while excluding air.

$$SnC_2O_4 = SnO + CO + CO_2.$$

It is a brownish-black powder, permanent in air at ordinary temperatures, but burning to dioxide if heated in air or oxygen.

The hydrate, $2SnO, H_2O$, is obtained as a white precipitate by addition of potash to stannous chloride. It dissolves in acids or in excess of potash, but not in ammonia.

Stannous Sulphide. SnS. Solution of stannous salts, such as the chloride, yield with hydrogen a brown precipitate of hydrated stannous sulphide. Like the oxide it dissolves in potash; it also dissolves in yellow ammonium sulphide, becoming by addition of sulphur converted into disulphide, SnS_2; when on adding acid to the solution, the disulphide is thrown down as a yellow precipitate.

Stannous solutions are converted into stannic compounds by oxidising agents, such as chlorine, bromine, aqua regia, etc.

Stannic Oxide. SnO_2. Tin dioxide. This compound is found native as tin stone or cassiterite, and is the oxide formed when metallic tin is heated in air. It is very insoluble in acids, but like silica, when fused with alkaline carbonates, forms a salt with the alkaline metal, which is soluble in water. Sodium stannate, for example, is Na_2SnO_3, corresponding to sodium carbonate Na_2CO_3, or sodium silicate, Na_2SiO_3.

Stannic Acid. $Sn(OH)_4$. When dilute acid is cautiously added to an aqueous solution of stannate, a white gelatinous precipitate of stannic acid is obtained of the composition $Sn(OH)_4$. By drying the precipitate in vacuo it loses water, becoming H_2SnO_3, and on application of strong heat it is rendered anhydrous.

Stannic acid dissolves in alkalies forming stannates, it is also soluble in hydrochloric acid.

A hydrated oxide of tin, known as metastannic acid, is produced by the action of nitric acid on metallic tin; it appears to be a condensed acid of the composition $H_{10}Sn_5O_{15}$.

Stannic Chloride. $SnCl_4$. The anhydrous chloride is obtained by passing chlorine gas over heated tin, or otherwise by the distillation of a mixture of tin and a sufficiency of corrosive sublimate:—

$$Sn + 2HgCl_2 = SnCl_4 + 2Hg.$$

It is a heavy fuming liquid, boiling at 116°C.; with a small quantity of water it forms a clear solution, but this on dilution decomposes into stannic acid, which separates as a white gelatinous precipitate, and soluble hydrochloric acid. Many double chlorides may be obtained by the union of stannic chloride with alkaline and other metallic chlorides; they are typified by the compound K_2SnCl_6. The vapour density of stannic chloride is approximately 130, and accordingly the formula is $SnCl_4$ (mol. wt. 260).

Stannic Sulphide. SnS_2. The disulphide of tin is used, under the name of 'Mosaic gold,' as a bronze powder. It is prepared by heating together an amalgam of tin and mercury with sulphur and sal ammoniac: some calomel and cinnabar sublime and a scaly mass of stannic sulphide remains. A yellow precipitate of stannic sulphide is obtained by the action of hydrogen sulphide on stannic solutions.

Like the dioxide, it possesses acid characters, it dissolves in potash and in alkaline sulphides, producing sulpho-stannates. Sodium sulpho-stannates of the composition Na_2SnS_3 and Na_4SnS_4 may be prepared in this manner, as yellowish or nearly colourless crystalline substances.

Lead. Pb. Plumbum. Atomic weight, 206. Specific gravity, 11·4.

The most important lead ore is **galena**, PbS, since this is almost the only source from which metallic lead is obtained.

The smelting of lead is very simple. After the ore has been separated from earthy matters it is placed on the hearth of a furnace, and heated in an oxidising flame, the air being freely admitted to the furnace. In this roasting sulphur burns off as dioxide, and a portion of the lead is converted into oxide, some sulphate also being formed. Then the mass is mixed together, the furnace closed, and the heat considerably increased. The metal is produced by the changes represented in the equations:—

$$\left.\begin{array}{l}2\,PbS + 3\,O_2 = 2\,PbO + 2\,SO_2\\ PbS + 2\,O_2 = PbSO_4\end{array}\right\}\text{Roasting.}$$
$$\left.\begin{array}{l}2\,PbO + PbS = 3\,Pb + SO_2\\ PbSO_4 + 2\,PbS = 3\,Pb + 2\,SO_2\end{array}\right\}\text{Reduction.}$$

In the reduction the lead sulphide and oxide react to form metallic lead and sulphur dioxide. In case of the oxidation being carried too far, the excess of oxide and sulphate of lead are reduced by an addition of carbon.

Refining. Lead containing foreign metals, such as tin or antimony, is purified by refining. For this purpose the metal is melted in shallow pans, and allowed to oxidise on the surface; the tin and other impurities are the first oxidised and can be removed by skimming.

Extraction of Silver. The refined lead still contains silver, which can be obtained in several ways. In the process devised by **Pattinson**, and named after him, the lead is melted in a large pot, one of a series of iron pots conveniently arranged in a row, and permitted to cool until **crystals** form in it. The crystals are nearly pure lead, and are ladled out from the more fusible silver-lead alloy as they form, and placed in the pot next on the right; about one-fifth of the metal is left at last, and this is removed to the pot next on the left. By a repetition of the process the silver is **concentrated** in the lead, which, when it reaches the last pot of the row, may contain 200 ounces to the ton (nearly 1 per cent.), while the desilverised lead will not contain more than half an ounce of silver to the ton. The quantity originally present may have been 5 or 6 ounces per ton.

Cupellation. To extract the silver from the rich

alloy, it is customary to separate all the lead as oxide. This operation is done on a **porous** hearth or **cupel**, made chiefly of bone ash (calcium phosphate), which forms the bed of a reverberatory furnace. The lead being melted, a current of air blown upon the surface converts it into the oxide, **litharge**, which fuses and runs off from the metal. At the end of the operation a brilliant mass of metallic silver is left. The lead converted into litharge is recovered by fusion with charcoal.

Another method of desilverising lead is the Parkes's process, in which zinc is used. A quantity of zinc is stirred into the melted lead, and after a time an alloy of zinc and silver rises to the surface and can be skimmed off. The silver is recovered from the alloy with zinc by distilling off the zinc. Any traces of zinc left in the lead, which, as lead does not take up more than a small limited quantity of zinc, are not large, are removed by exposing the melted metal to the oxidising action of the air, or a current of superheated steam, whereby the zinc becomes converted into oxide, and pure lead left.

Lead is a soft, white metal; a freshly-scraped surface being lustrous, but rapidly tarnishing and becoming bluish when exposed to the air. It melts at 330°, and is volatilised at high temperatures. If a stick of zinc be hung in a solution of lead acetate (sugar of lead), the metallic lead will deposit on the zinc and grow into a mass of **crystalline** plates, known as a **lead tree**. Lead is drawn into pipes for carrying water, and in the form of sheet-lead is used for roofing. It is a poisonous metal, and taken even in small quantities accumulates in the body, eventually exerting a poisonous action.

Lead alloyed with tin forms pewter or solder; with antimony the alloy used for casting **printing type** is obtained. When used for making shot a little arsenic is commonly added to the lead to harden it.

Metallic lead dissolves in nitric acid, but not in hydrochloric acid or aqua regia, except in limited quantity, the chloride of lead being insoluble; sulphuric acid is almost without action on it.

In its compounds lead commonly appears as a dyad, as, for example, in PbO, $PbCl_2$, $PbSO_4$; but in some compounds it shows tetrad characters, as in PbO_2, and especially in the volatile organic compound lead tetrethide, $Pb(C_2H_5)_4$; in which an atom of lead is combined with four monovalent ethyl (C_2H_5) groups.

The following are some of the chief lead compounds:—

PbO	Lead protoxide.	PbO_2	Lead dioxide.
Pb_2O_3	Lead sesquioxide.	$PbCl_2$	Lead chloride.
Pb_3O_4	Red lead oxide.	PbN_2O_6	Lead nitrate.

Lead Protoxide. PbO. Litharge.

Lead protoxide is formed when lead is heated in air; at a high temperature a scaly mass of **litharge** is obtained, but at a lower heat the *yellow* powder known as **massicot** is formed. When alkalies are added to lead solutions, such as nitrate or acetate, a *white* precipitate of lead hydrate is thrown down; which when dried and heated turns yellow, as it loses water.

This lead oxide is a basic substance, with slightly alkaline characters; it absorbs carbonic acid from the air; and is soluble in the alkalies potash and soda.

Red Lead Oxide. Pb_3O_4. The oxide known as **Red Lead** is somewhat variable in composition, some specimens being composed chiefly of the sesquioxide Pb_2O_3. Red lead is made by exposing the powdered protoxide or massicot in a furnace at a temperature of $320°$ C., oxygen is then absorbed.

$$6PbO + O_2 = 2Pb_3O_4.$$

Lead Salts.

It has the character of a compound oxide, as nitric acid removes protoxide from it, leaving dioxide, thus:—

$$Pb_3O_4 + 4HNO_3 = 2PbN_2O_6 + PbO_2 + 2H_2O.$$

Red lead is used as an oxidising substance, especially in the manufacture of flint glass.

Lead Dioxide. PbO_2. Peroxide of Lead.

This oxide is left as a brown powder by the action of nitric acid on red lead. It does not dissolve in that acid, and if heated with sulphuric acid forms sulphate, oxygen being liberated. It is soluble in strong hydrochloric acid, forming a yellow liquid, which appears to contain a tetrachloride, $PbCl_4$. Lead dioxide is otherwise obtained by adding alkaline hypochlorites to lead solutions.

Lead Carbonate. $PbCO_3$. White Lead.

The normal carbonate of lead is found as a mineral, and may be obtained by precipitation; a basic compound, or hydrate-carbonate, is manufactured on a large scale, and commonly called **White Lead**. In the 'Dutch process' of manufacture, small gratings of cast lead are exposed to the vapour of vinegar and carbonic acid. The vinegar (acetic acid) being placed in pots, and lead placed above it; the pots are buried in spent tan or manure. By the joint action of the acid and oxygen of the air a basic lead acetate is produced, which in turn becomes slowly changed into carbonate by the carbonic acid gas evolved in the fermentation of the vegetable matter. The acetic acid thus liberated produces fresh acetate, and only a small quantity of the acid is required for the operation. After some weeks the metal becomes entirely converted into a white crust of carbonate; of which the composition is commonly $2PbCO_3, Pb(OH)_2$.

Lead Nitrate. PbN_2O_6. The nitrate is easily prepared by the action of dilute nitric acid upon lead, or by dissolving litharge in nitric acid; it is a white salt, soluble in water, and crystallising in octahedra. When strongly heated it decomposes into lead oxide, nitrogen peroxide, and free oxygen.

$$PbN_2O_6 = PbO + 2NO_2 + O.$$

A solution of lead nitrate with hydrochloric acid or a soluble chloride, yields a white precipitate of lead chloride, $PbCl_2$.

CHAPTER XXVI.

PENTAD METALS.

Antimony and Bismuth.

ANTIMONY and BISMUTH belong to the pentad group of elements, which form the following complete series: Nitrogen, Phosphorus, Arsenic, Vanadium, Antimony, and Bismuth.

Antimony. Sb. Stibium. Atomic weight, 122. Specific gravity, 6·8.

The metal antimony is extracted from the mineral sulphide, or antimonite, Sb_2S_3. The sulphide being fusible, is separated from earthy matters by melting, and reduced in a crucible by charcoal and sodium carbonate. The alkali forms sodium sulphide, and the antimony becomes changed into oxide, which latter by the reducing action of the carbon is converted into metal.

$$\begin{cases} Sb_2S_3 + 3Na_2CO_3 = Sb_2O_3 + 3Na_2S + 3CO_2. \\ Sb_2O_3 + 3C = Sb_2 + 3CO. \end{cases}$$

Antimony is a bright, white metal, brittle, and crystalline in structure. It fuses at a temperature of 450° and volatilises at a red heat. The metal is permanent in air when cold; but rapidly oxidises when heated above melting-point. It combines with chlorine eagerly, and the finely-divided metal takes fire if sprinkled into a jar of the gas.

The compounds of antimony fall into two groups of

antimonious and antimonic compounds, of which the following are the more important:—

Antimonious compounds.

$SbCl_3$ Antimony trichloride.
SbH_3 „ hydride.
Sb_2O_3 „ trioxide.
$SbOCl$ „ oxychloride.
Sb_2S_3 „ trisulphide.

Antimonic compounds.

$SbCl_5$ Antimony pentachloride.
Sb_2O_5 Antimony pentoxide.
Sb_2O_4 or [Sb_2O_3, Sb_2O_5] Antimony tetroxide.

Antimony Trichloride. $SbCl_3$. This chloride is prepared by acting on excess of antimony with chlorine; by dissolving antimony sulphide in strong hydrochloric acid and evaporating, or by dissolving the metal in aqua regia. If the evaporation be done in a retort, after the acid has been driven off the trichloride begins to distil over and condenses as a white crystalline solid, known as butter of antimony. It fuses at 72°C., boils at 223°, and the density of the vapour is 114, corresponding to the molecule $SbCl_3$.

By the action of water it is converted into the white insoluble oxychloride:—

$$SbCl_3 + H_2O = SbOCl + 2HCl.$$

Antimony Pentachloride. $SbCl_5$. This chloride, corresponding to phosphorus pentachloride, can only be obtained by acting on the metal with chlorine. It is a fuming liquid, and decomposes with water into (1) the oxychloride, $SbOCl_3$; (2) the oxychloride, SbO_2Cl; and (3) antimonic acid, $SbO_2(OH)$.

The oxychlorides and the antimonic acid are obtained as white precipitates.

Antimony pentachloride separates, when heated, into trichloride and free chlorine.

Antimony Trioxide. Sb_2O_3. If metallic antimony be heated in a tube in a slow current of air crystals of the trioxide are obtained, which pass with access of more oxygen into the tetroxide Sb_2O_4. The trioxide is thrown down as a white precipitate on pouring solution of trichloride into boiling sodium carbonate. It is a whitish powder, soluble in caustic potash, forming the unstable antimonite, $KSbO_2$; but its acid properties are feeble.

If antimony trioxide is boiled with a solution of acid potassium tartrate, it dissolves, and on crystallising, the salt known as Tartar Emetic is obtained.

$$Sb_2O_3 + 2KHC_4H_4O_6 = 2K(SbO)C_4H_4O_6 + H_2O.$$

Antimony Tetroxide. Sb_2O_4. This is the most stable of the antimony oxides: it is obtained by heating the trioxide in air, or by heating the pentoxide to redness, when it loses oxygen. It is also prepared by acting on metallic antimony with strong nitric acid, and heating the white residue of antimonic acid so produced. It is a heavy infusible powder, and hardly soluble in acids. With acid potassium tartrate it decomposes, forming tartar emetic, and leaving antimonic acid.

$$(SbO)SbO_3 + KHC_4H_4O_6 = K(SbO)C_4H_4O_6 + HSbO_3.$$

Antimony Pentoxide. Sb_2O_5. The pentoxide of antimony is left as a white powder when antimonic acid is gently heated.

Antimonic Acids.

There are three typical antimonic hydrates corresponding with the phosphoric acids, but not quite as distinct from each other in characters.

$$Sb_2O_5 + 3H_2O = 2H_3SbO_4,\ \text{ortho-antimonic acid.}$$
$$Sb_2O_5 + 2H_2O = H_4Sb_2O_7,\ \text{metantimonic acid.}$$
$$Sb_2O_5 + H_2O = 2HSbO_3,\ \text{antimonic acid.}$$

The first hydrate (ortho-antimonic) is unknown, and

no salts have been produced corresponding with it; but potassium and other metallic antimonates may be obtained of the composition $K_4Sb_2O_7$ and $KSbO_3$.

When metallic antimony is treated with strong nitric acid, containing some hydrochloric acid, a white almost insoluble powder of antimonic acid, $HSbO_3$, is obtained and an acid $HSbO_3\,2H_2O$ or H_5SbO_5 is formed by acting on potassium antimonate with nitric acid.

Antimony Sulphides. Antimony trisulphide (hydrated) is obtained as an orange precipitate by the action of hydrogen sulphide on antimonious solutions. If dried and heated it fuses, and becomes grey and crystalline like the native sulphide. It may be also made directly by heating antimony with sulphur. Antimony sulphide is a feebly acid substance; it dissolves in caustic alkali, and also in solutions of alkaline sulphide, producing **Sulphantimonites**; meta-sulphantimonites being typified by—

$$Na_2S + Sb_2S_3 = 2NaSbS_2;$$

and ortho-sulphantimonites by—

$$3Na_2S + Sb_2S_3 = 2Na_3SbS_3.$$

Antimony trisulphide, when boiled with free sulphur and sodium sulphide, forms a soluble sulphantimonate, thus:—

$$Sb_2S_3 + S_2 + 3Na_2S = 2Na_3SbS_4.$$

This sodium sulphantimonate is known as Schlippe's Salt; it can be obtained in crystals by evaporating the liquid. If treated with acids it breaks up, and hydrated antimony pentasulphide is obtained as an orange precipitate.

Antimony Hydride. SbH_3. **Antimonuretted Hydrogen** or **Stibine.** The hydride of antimony is formed under similar conditions in the same manner as arsenic hydride. Its production may be shown with the apparatus used for **Marsh's Test for Arsenic**. Zinc and dilute sulphuric acid are placed in a flask, and solution of antimony chloride added by means of the thistle funnel: antimony hydride comes off mixed with hydrogen. When the gas is burnt white fumes of trioxide are produced, and black stains are deposited upon porcelain brought into the flame. The metal is deposited as a dark mirror if the gas be passed through a heated tube. Antimony stains differ from those of arsenic and can be distinguished by their black colour, by not dissolving in hypochlorite solutions, and by the difference in volatility. Stibine has never been obtained quite pure and unmixed with hydrogen, see page 148.

Bismuth. Bi. Atomic weight, 209. Specific gravity, 9·8.

Bismuth is a somewhat rare metal, commonly found native, but sometimes as oxide, Bi_2O_3, or sulphide, Bi_2S_3. The ore is roasted, and afterwards reduced by carbon with the addition of a flux; as the metal melts at a very low temperature it easily separates from infusible impurities.

Metallic bismuth is a brittle metal, of a bright, almost white colour, but with a slight tinge of red or rose colour. It forms very distinct crystals in cooling from the melted state. When pure it melts at about 270° C., but alloyed with lead it forms very fusible alloys. Rose's fusible metal (m. p. 94°), which melts in boiling water contains two parts of bismuth, one of lead and one of tin. Such fusible alloys are used for casting letter press, engraved blocks, etc., and for making safety plugs for steam boilers.

Bismuth dissolves readily in nitric acid and aqua regia; it is indifferent to hydrochloric acid and cold sulphuric acid; but concentrated sulphuric, when heated, produces bismuth sulphate. The compounds of bismuth resemble those of antimony, but are more marked in their basic than in their acid characters.

The following will serve as examples:—

Bi_2O_3	Bismuth	trioxide.
$BiCl_3$	,,	trichloride.
$Bi(NO_3)_3$	,,	trinitrate.
BiH_2NO_5	,,	basic nitrate.
$BiOCl$	,,	oxychloride.
$HBiO_3$	Bismuthic acid.	
Bi_2O_5	Bismuth pentoxide.	

Bismuth Nitrate. BiN_3O_9, or $Bi(NO_3)_3$.

The salt is obtained in good crystals by dissolving the metal in nitric acid and evaporating the solution. The crystallised salt contains $Bi(NO_3)_3, 3H_2O$ or $BiH_6N_3O_{12}$. It is strongly acid in character, soluble in nitric acid, but decomposed by water into basic nitrate and free nitric acid.

Bismuth Oxynitrate or Basic Nitrate. BiH_2NO_5, or $(BiO)NO_3.H_2O$. This nitrate of bismuth is obtained as above described from the trinitrate. The metal is dissolved in nitric acid, the solution concentrated to expel excess of acid, and then poured into 80 to 100 volumes of cold water.

$$Bi(NO_3)_3 + 2H_2O = BiH_2NO_5 + 2HNO_3.$$

The basic nitrate is precipitated as a white powder, which is used in pharmacy, being known as *Bismuthi Subnitras*, or subnitrate of bismuth.

Bismuth Trichloride. $BiCl_3$. The highest chloride of bismuth is the trichloride: it is formed by the direct combination of the metal and chlorine, or in solution by acting upon the trioxide with hydrochloric acid. The solution may be evaporated to expel the excess of acid and the resulting white salt purified by sublimation or distillation.

Bismuth trichloride is a white crystalline substance, decomposed by water, but dissolving in hydrochloric acid. If the solution of the chloride is poured into water a white precipitate of **oxychloride** is produced:—

$$BiCl_3 + H_2O = BiOCl + 2HCl.$$

Bismuth Trioxide, Bi_2O_3. This trioxide is obtained by heating the hydrate, carbonate, or nitrate. When a solution of bismuth is decomposed by cold potash the hydrate is formed as a white precipitate, $Bi(OH)_3$; but if the liquids are boiling the trioxide is at once produced. It does not dissolve in alkalies, and is almost devoid of acid characters.

Bismuthic Acid. $HBiO_3$. This compound is formed by passing chlorine gas into boiling solution of potash containing bismuth trioxide in powder.

$$Bi_2O_3 + 2Cl_2 + 4KOH = 2HBiO_3 + 4KCl + H_2O.$$

This is practically equal to producing the oxidation by potassium hypochlorite. Bismuthic acid is a feeble acid, and easily decomposed on heating first into bismuth pentoxide and further into trioxide and oxygen.

$$2HBiO_3 = Bi_2O_5 + H_2O.$$

There is a striking difference in activity between this bismuth acid, $HBiO_3$, and the powerful nitrogen acid, HNO_3, the first of the pentad series of elements.

CHAPTER XXVII.

Atomic Weights in Relation to the Specific Heats of Elements. Atomic Heat.

By the **specific heat** of any substance is meant the amount of heat required to raise one gramme of the substance through one degree of temperature; and this value is commonly expressed in relation to the specific heat of water as unity. If our **thermal unit** be the heat required to raise the temperature of a gramme of water from 0° to 1°C., we can define the specific heat of any substance as being the number of thermal units required to raise one gramme through one degree in temperature. This value has been found for a large number of bodies, for compound bodies as well as for most of the elements, some of the results being given in the table opposite. In the first column are the elements, in the second the specific heats, in the third the atomic weights, and in the fourth the products of the specific heat multiplied by the atomic weight. Since the specific heat refers to the heat used or required by one gramme, if this value be multiplied by the atomic weight, we get the quantity of heat used by the atom in each case: which is described as the **atomic heat.**

$$\text{Sp. heat} \times \text{at. weight} = \text{atomic heat.}$$

Now it will be seen that while in the table the specific

heats form a descending series, the atomic weights are an ascending series: the metal Lithium having the lowest atomic weight, but the highest specific heat, while Uranium has the highest atomic weight and the lowest specific heat. The products obtained by multiplying them together are all approximately equal; that is, the **atomic heats of these elements are all substantially equal to each other.**

TABLE OF SPECIFIC HEATS, ETC., OF METALLIC ELEMENTS.

Element.		Specific Heat.	Atomic Weight.	Atomic Heat.
Lithium	Li	·941	7	6·6
Sodium	Na	·293	23	6·7
Magnesium	Mg	·250	24	6·0
Aluminium	Al	·225	27	6·1
Potassium	K	·166	39	6·5
Iron	Fe	·114	56	6·3
Nickel	Ni	·108	58·6	6·3
Cobalt	Co	·107	59	6·3
Copper	Cu	·095	63	6·0
Zinc	Zn	·093	65	6·1
Arsenic	As	·082	75	6·1
Silver	Ag	·056	108	6·0
Cadmium	Cd	·055	112	6·1
Tin	Sn	·055	118	6·5
Antimony	Sb	·050	120	6·0
Platinum	Pt	·032	195	6·3
Gold	Au	·032	196	6·3
Mercury	Hg	·032	200	6·4
Lead	Pb	·031	206	6·4
Bismuth	Bi	·031	209	6·4
Uranium	U	·028	240	6·6

This generalisation was first made by **Dulong and Petit**, and they contended that the law would be found to apply to all elements. As far as the metallic elements are concerned the agreement is fairly good, and the atomic heats are always about 6·4. In the following table the specific heats of some non-metallic elements are given, which it will be seen have atomic heats differing greatly from each other.

TABLE OF SPECIFIC HEATS, ETC., OF NON-METALLIC ELEMENTS.

		Specific Heat.	Atomic Weight.	Atomic Heat.
Boron	B	·250	11	2·7
do. at 1000° C.		·500	11	5·5
Carbon diamond	C	·147	12	1·7
do. do. at 1000° C.		·460	12	5·5
graphite		·198	12	2·4
Silicon	Si	·166	28	4·6
do. above 230°		·203	28	5·6
Phosphorus crystal	P	·174	31	5·4
amorphous		·170	31	5·3
Sulphur rhombic	S	·178	32	5·7
Chlorine gas	Cl	·093	35·5	3·3
solid		·180	35·5	6·6
Bromine liquid	Br	·111	80·	8·8
solid		·084	80·	6·7
Iodine	I	·054	127	6·8

It is evident from these facts that the atomic heats of some elements are far removed from the value 6·4. The difference is most marked in the cases of the elements Boron, Carbon, and Silicon, for which the determinations of specific heats were made in the ordinary way at temperatures below 100° C. But it should be recognised that while the atomic weight of an element is an unvarying ratio, it is found that the specific heat of an element varies, in some instances considerably, according to its physical state, being greatest for the solid and least for the gaseous form, and also it increases with a rise of temperature. At high temperatures the atomic heats given in the table for carbon, boron, and silicon are approximately 5·5, which is nearly the same value as that given for phosphorus and sulphur at ordinary temperatures. These elements of course be-

Molecular Heats. 251

come gaseous at high temperatures, while the former three do not, and no comparisons at extreme temperatures are possible.

By the application of the law of Dulong and Petit we may be able, in doubtful cases, to decide the atomic weight of a metal. For example, the metal Indium combines with chlorine in the proportion of $57 : 35.5$, and we might write the formula InCl (atomic weight of indium 57.0), or we might suppose the chloride to be $InCl_2$ (atomic weight 114), or even $InCl_3$ (atomic weight 171); in each case the ratio of metal to chlorine would be the same. But when it is found by a direct experiment that the specific heat of Indium is $.057$ we cannot hesitate to accept 114 as the atomic weight, since ($114 \times .057 = 6.5$) this value gives 6.5 as the atomic heat of the metal.

Molecular Heats of Compounds.

The **molecular heat** of a compound is found in the same manner as the atomic heat of an element, by multiplying together the specific heat and molecular weight.

For example, potassium chloride has the specific heat 1.73 and molecular weight 74.5, whence ($.173 \times 74.5 = 12.8$) the molecular heat is 12.8. It is found that the molecular heats of the chlorides of the type MCl average 12.8, those of the type MCl_2 are about 18.5.

It has been shown in numerous instances that the molecular heat of a compound is equal to the sum of the atomic heats of the constituent elements in the solid form; and it is probable that with a better knowledge of the true specific heats of the bodies in question the generalisation will be more perfectly established. Thus the sulphides of the formula MS have a specific heat of about 12 and $6.4 + 5.7 = 12.1$. If we assume the truth of this relation we have a means of estimating approximately the specific heats of solid oxygen, nitrogen, and hydrogen which cannot be determined by any direct experiment.

Thus the molecular heats of—

$$KCl = 12\cdot 8$$
$$KClO_3 = 24\cdot 8$$
$$O_3 \text{ by difference} = 12$$

and the atomic heat of oxygen is 4,

$$KNO_3 = 24\cdot 1$$
$$K = 6\cdot 4$$
$$O_3 = 12$$
$$\text{by difference } N = 5\cdot 7$$

or about the same value as for phosphorus,

$$H_2O \text{ (ice)} = 9$$
$$O = 4$$
$$H_2 \text{ by difference} = 5$$

or the atomic heat of solid hydrogen is approximately 2·5.

CHAPTER XXVIII.

THE PHYSICAL PROPERTIES OF GASES.

The three States of Matter, viz.: the **solid** state, the **liquid** state and the **gaseous** state depend upon certain conditions of temperature and pressure. We know, for example, that some solids when heated melt and become liquid, and some liquids when heated boil and are converted into **vapours**; as ice becomes water and water steam. Although there is no real distinction between a vapour and a gas, the term vapour is used generally in respect of those gases which easily condense under ordinary conditions of temperature, as distinguished from those which remain gaseous under the same conditions. But since all gases are now known to be condensible, all alike are to be regarded as vapours produced from liquids; some indeed boiling at extremely low temperatures and others at very high temperatures.

The changes of state which matter undergoes are, in the first place dependent upon the action of heat, but the conditions as regards pressure are also of great importance. As regards the fusion of solids, the melting-point is affected, but only to a small extent, by changes of pressure; but the temperature at which a liquid boils or is entirely converted into vapour, *the boiling point*; and *vice versa* the temperature at which a vapour condenses to a liquid are, as we shall see, entirely governed by the pressure at the moment of the change.

Solid bodies exhibit infinite degrees and varieties of

cohesion, and the terms hard, brittle, elastic, tenacious, ductile, malleable, and the like, are used to express the resistance to change of form possessed by such bodies. In solids, in other words, it is evidently difficult by using force or pressure to cause the particles or molecules to move upon each other, but the difficulty is greater in some cases than in others. When we apply heat to a solid, since the solid expands, obviously the molecules are set in motion, but within limits a change of size is the only noticeable result. On further heating, we may produce a chemical change or decomposition, or the body melts to a liquid, the change of state being accompanied by an absorption of heat.

Liquids possess slight cohesion, being easily poured and divided, and the molecules move readily about each other, so that liquids take the form of the containing vessel with a horizontal free surface; left to themselves they form spheres or drops. The mobility of liquids or their **viscous** characters still are measurably different; thus in very mobile liquids, such as ether or alcohol, bubbles formed by shaking rise and break quickly, much more rapidly indeed than they do in water, while oil, glycerine, solutions of sugar, gum, and soap, pour slowly and retain bubbles for some time.

The free surface of any liquid, by reason of the cohesion between the molecules, is in a condition of tension, and requires a perceptible application of force to break it.

Latent heat of fusion. The change of state from the solid to the liquid, whether by simple fusion or in the act of dissolving a solid in water or other liquid, is accompanied by a loss of heat. The heat thus absorbed or rendered latent when ice is melted is about 80 c., see p. 37.

Gases. 255

Gases or vapours are produced by the continued heating of liquids; they expand on heating, the rates of expansion being generally larger than in the case of solid bodies, and also increasing rapidly as the temperature rises. When we are dealing with a volatile liquid, which does not suffer chemical division, vapour is continually given off from the surface as the temperature rises until the *boiling point* is reached, when the tension of the vapour becomes equal to the atmospheric pressure.

To take the behaviour of water as an example: suppose a few drops of water are passed into a barometer tube at $4°$ C., say with the mercury at 760 m.m., we find that the column of mercury sinks at once to 754 m.m., giving 6 m.m. as the **vapour tension** of water at $4°C$. If the tube be gradually heated the vapour tension increases, as is shown by the continued sinking of the mercury, until at $100°C$. it becomes equal to the atmospheric pressure, and the mercury stands at the same level inside and outside of the tube. This increase is shown in the following table :—

Vapour Tensions of Water.

Temperature.	Tension in millimetres.
$0°$	4·6
10	9·1
20	17·4
40	54·9
60	148·8
80	354·6
90	525·0
100	760·0

The vapour under the conditions of the experiment is a **saturated** vapour and any increase of pressure produces liquefaction. The vapour tensions of other liquids have been determined in a similar manner.

The boiling point of water under barometric pressure

of 760 m.m. is 100°C., for at this temperature the tension of water vapour is equal to the atmospheric pressure. If we lower the pressure the boiling point is altered, thus under a pressure of 525 m.m. water boils at 90°, or under pressure of 354 at 80°, and so on as in the table.

Latent heat of vaporization. The conversion of a liquid into a vapour or gas is accompanied by a striking increase of volume; thus one cubic centimeter of water at 4° which weighs one gramme, only expands by heating to 100° by about one twentieth part, becoming 1·05 c.c.; in the form of steam however it measures more than 1200 c.c. By the action of heat the molecules of water are thus driven far apart and set in active motion in all directions; and so long as the heating is maintained this free motion of the gaseous molecules continues. This great expansion is accompanied by absorption of heat, used in the work of driving the molecules apart, which since it no longer affects the thermometer is described as *latent heat*. The method for finding the quantity of heat, appearing as *sensible* heat, when steam is passed into water is described on p. 37; the value thus obtained is 563 c.

Liquefaction of Gases. The inversion of the heating or vaporising process we have described produces liquefaction, and it is now demonstrated that all gases can be reduced to the liquid state. In order to effect this, however, it is necessary to employ in many cases an increased pressure as well as a low temperature.

Sulphur dioxide is liquefied readily by a freezing mixture or by a pressure of 1·5 atmospheres; Ammonia requires 4·5 atmospheres at 0°C., Chlorine 9 atmospheres, and Hydrogen Sulphide 10 atmospheres.

The liquefaction of ammonia in a sealed glass tube was shown by Faraday in the following way:—A well annealed strong glass tube, bent as in figure 55 with one end sealed, is charged with a solid compound of silver

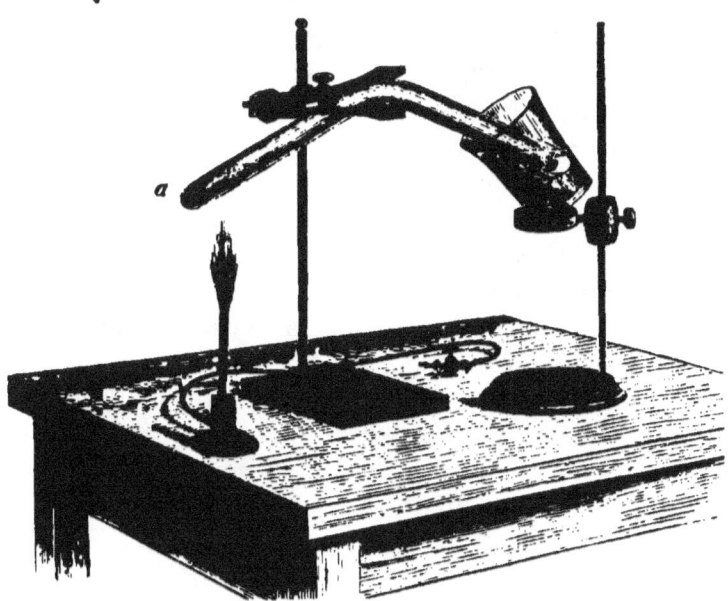

Fig. 55.

chloride, or calcium chloride, saturated with ammonia gas: the other end is then carefully sealed in the blowpipe. Then if the end a be heated while the end b is cooled with a freezing mixture, the ammonia will be given off as gas from the compound, and the pressure will increase within the tube until the ammonia liquefies and condenses in the cold part of the tube.

Liquefied Nitrous oxide and Carbon dioxide can be obtained by pumping the gases into strong iron cylinders with powerful condensing pumps; the pressures or vapour tensions of the condensed gases are about 33 and 40 atmospheres respectively.

The gases most difficult to liquefy, and which have in

s

fact only within a few years been condensed to the liquid state, are Hydrogen, Oxygen, Nitrogen; and the compound gases Nitric oxide, Carbonic oxide, and Marsh gas. Oxygen was first liquefied in 1877, and quite independently by different methods by two experimentalists, Messrs. Pictet and Cailletet. Since that date all the remaining gases have proved to be liquefiable. In order to liquefy oxygen, a very low temperature is obtained in the condensing tube, by the evaporation of condensed carbon dioxide or liquefied ethylene (C_2H_4); the latter gas under a pressure of 1 m.m. boiling at -162 C.; at this low temperature oxygen condenses to a liquid by a pressure of about 5 atmospheres. When the pressure is taken off the liquid oxygen boils in the tube producing still lower temperatures, and by the intense degree of cold thus obtained carbonic oxide and nitrogen have been frozen to snow-like solids.

Liquefied air under a reduced pressure of 10 m.m. boils at $-220°$C.

The following table gives the temperatures at which the several liquids boil; under a pressure of 760 m.m. of mercury:—

Boiling Points.

	Temp.	Press.
Sulphur dioxide	$-10°$ C.	760 mm.
Ammonia	$-33°$	,,
Carbon dioxide	$-78°$	,,
Nitrous oxide	$-105°$	,,
Nitric oxide	$-154°$	,,
Marsh gas	$-164°$	,,
Carbonic oxide	$-190°$	,,
Oxygen	$-182°$	,,
Nitrogen	$-194°$	,,
Hydrogen (about)	$-200°$	,,

The low temperatures are measured either by the contraction of hydrogen in a suitable glass vessel or by a thermoelectric couple of copper and german silver used with a galvanometer.

Critical temperature. If heat be applied to a liquid contained in a strong sealed glass tube so that the vapour cannot escape, when the temperature rises to a given point the liquid loses its free surface, and no difference between the liquid and its vapour can be detected; the tube being apparently full of gas. It is evident that during the heating the density of the liquid diminishes by expansion while the density of the vapour continually increases with the increase of pressure; so that at length the densities of liquid and gas are equal. At the instant when this point is reached the free surface flickers and disappears, and the liquid state becomes continuous with the gaseous state. The temperature at which the change occurs is termed the **critical temperature**; in the case of water the critical temperature is reached at about 360°C.; for alcohol the temperature is 207°, for sulphur dioxide 210°, and for carbon dioxide 30·9°. Above its critical temperature a gas cannot be reduced to a liquid by any pressure however great. Thus, at $-118°$C. oxygen may be liquefied by a pressure of 50 atmospheres, but above such a temperature no visible liquefaction can be obtained, although the compressed gas may become more dense than liquid oxygen itself.

Diffusion of gases. A gas when contained in a vessel fills it so as to come in contact with the inside surface everywhere; it differs thus from a liquid which only fills that part of the vessel containing it and has a free surface. From the free surface of a liquid a continual escape of molecules of the liquid may take place by evaporation, but by the attraction of molecules for each other the cohesion is sufficient to renew and maintain the surface of a liquid. This is not the cases with gases,

Diffusion of Gases.

and if the vessel containing any gas be open to the air at any point the gas will escape outwards while air will pass inwards.

Or, similarly, if we bring two vessels containing different gases into connection with each other, a perfect mixture of the gases by diffusion into each other will take place.

This property of gaseous diffusion may be shown with two flasks connected by a tube, as in fig. 56. The upper flask is filled with hydrogen and the lower with the heavier oxygen; after a time an explosive mixture will be found in each flask, some hydrogen having travelled downwards while oxygen has passed up.

The diffusion of gases will also take place through porous substances, such as plaster of Paris, unglazed earthenware, or discs of compressed plumbago, blotting paper, etc. Hydrogen diffuses with remarkable rapidity, as may be shown by the apparatus in fig. 57.

Fig. 56.

The jar b, in the figure, is closed with a porous disc which is prepared by mixing plaster of Paris to a cream with water in a plate and standing the jar in it: when the plaster sets it forms a porous bottom to the jar. The plaster is allowed to dry, and being covered with a glass plate the jar is filled with hydrogen: if the mouth of the tube be brought into a little coloured water, and the plate removed, the hydrogen diffuses

Fig. 57.

rapidly out of the jar, and air passes in less rapidly, so that the water is sucked up the tube into the jar.

The rates of diffusion of different gases were the subject of a number of careful experiments by Graham, who observed the times required for the passage of different gases, either through tubes or minute holes, or through porous plates of such materials as graphite, plaster of Paris, or earthenware. It was found that *the rate of diffusion is in inverse ratio to the square root of the density* of the gas, and this relation holds true whether the diffusion takes place into another gas or into a vacuum. The following table will show how nearly the rate thus calculated agrees with the results of experiment; the velocities of diffusion are referred to air as the standard of both density and velocity of diffusion :—

Gas.	Density.	Square root of density.	$\dfrac{1}{\sqrt{\text{density}}}$	Velocity of diffusion.
Air	1	1	1	1
Hydrogen	·0692	·263	3·779	3·83
Marsh gas	·559	·747	1·337	1·344
Nitrogen	·971	·985	1·014	1·014
Oxygen	1·105	1·051	·951	·948
Carbon dioxide	1·529	1·236	·808	·812

It will be seen from the table that the velocity of diffusion of hydrogen is four times that of oxygen, while the square roots of their relative densities are in the ratio of one to four.

Expansion of gases by heat.

The general effect of heating a gas is to increase the volume if the pressure remains unchanged. We can easily show the fact of expansion of air when heated by the following experiment: a small flask is fitted with a cork and tube like the upper flask b in figure 56, and supported with the end of the tube dipping into water. If the flask be warmed with a flame, or by pouring hot water over it, the air within becomes warmer, and in expanding some is driven out and escapes in bubbles through the water; and the pressure in the flask is of course unaltered. Then, as the flask cools again, the air contracts and a certain quantity of water will pass up the tube and possibly into the flask.

To express definitely the amount of expansion of a gas it is necessary to compare the measured expansion with the volume at a definite point, and the standard selected as convenient is the volume at the temperature of melting ice or 0°C. By careful experiments it has been found that one volume of air at 0°C. when heated to 100° becomes 1·366, and that other gases expand by nearly the same amount. From this it is plain that (the expansion being uniform) the increase of volume for one degree is ·00366 of the *volume at zero*, and this factor which is very nearly represented by $\frac{1}{273}$, is termed the *coefficient* of expansion.

The law is expressed conveniently by the formula—

$$V_t = V_0 + V_0 \times \frac{t}{273},$$

where V_t is the volume at $t°$ and V_0 the volume at 0°, from which it follows that—

$$V_t = V_0 \frac{273+t}{273} \quad \text{and} \quad V_0 = V_t \frac{273}{273+t}.$$

Since the expansion is $\frac{1}{273}$ of the volume at 0°, then 273 c.c. at 0° become 274 c.c. at 1°, 275 c.c. at 2°, and so on.

We can illustrate the use of the formula by two examples:—

Suppose a quantity of hydrogen to measure 250 c.c. at 0°, what will it measure at 60°?

$$250 \times \frac{273+60}{273} = \text{volume at } 60°,$$

$$250 \times \frac{333}{273} = 304·9 \text{ c.c.}$$

50 c.c. of air are measured at 17°C.; find the volume at 0°:

$$V_0 = 50 \times \frac{273}{290} = 47·07 \text{ c.c.}$$

A simpler formula for calculation is—

$$\frac{V}{V_1} = \frac{T}{T_1},$$

in which V and V_1, are the volumes and T and T_1 are the absolute temperatures obtained by adding 273 to the observed temperatures t and t; viz.—

$$T = 273 + t \text{ and } T_1 = 273 + t_1.$$

This formula is equivalent to the statement that the volume of a gas varies directly as the absolute temperature.

Boyle's or Marriotte's law. We have now to consider the effect of **changes of pressure** on the **volume of a gas**, the temperature being unchanged.

This may be shown with a U-shaped bent tube, with its shorter limb closed and the longer leg open to the air; some mercury is poured down the longer limb so as to enclose the air in the shorter limb, and when the mercury is level in the two tubes the enclosed air is obviously at atmospheric pressure, say at 760 m.m. If more mercury is then poured in the air will be compressed into a smaller space, the pressure upon it being indicated by the difference of level of the mercury. When the mercury in the long leg stands at a height of 760 m.m. (or about 30 inches) above that in the short leg, the volume will be one half of the original volume and the total pressure is the original atmospheric pressure of 760 m.m., plus the added mercurial column of 760 m.m. in high or two atmospheres. Thus by doubling the pressure the volume is halved. If the pressure be increased by another atmosphere the volume becomes one-third, and so on. It has been demonstrated by accurate measurements that for air and most gases the volume of the gas varies *inversely* as the pressure. The law thus stated is commonly known as Boyle's law.

We may conveniently express this fact by the formula—

$$\frac{P}{P_1} = \frac{V_1}{V},$$

where P and P_1 are the pressures and V and V_1 the volumes. From this it follows that $P \times V = P_1 \times V_1$ or PV is constant, and also $V_1 = V\dfrac{P}{P_1}$, from which we can calculate the volume at any pressure.

To take an example: a gas in a eudiometer, standing over mercury, measures 96 c.c. at a pressure of 350 m.m., required the volume at 760 m.m. pressure:—

$$V_1 = 96 \times \frac{350}{760}; \quad \therefore \quad V_1 = 44\cdot2 \text{ c.c.}$$

It should be observed that Boyle's law is approximately correct for all gases when far removed from their liquefying point, but at high pressures, as a gas approaches the conditions of its change to the liquid state, the relation no longer holds good. It is, however, used for all ordinary purposes of calculation in regard to gases, such as arise in the analysis by the eudiometer and similar problems.

A flask holding 4 litres is filled with oxygen at 150°C, the barometer standing at 750 m.m., required the volume of oxygen at 0° and 760 m.m. Here we must reduce both temperature and pressure to the standard; the corrected volume is—

$$4000 \times \frac{750}{760} \times \frac{273}{288}; \text{ or } 3742 \text{ c.c.}$$

Suppose the weight of the gas to be required: one gramme of hydrogen measures 11·2 litres (11200 c.c.) at 0° and 760 m.m., and the density of oxygen being 16, it follows that 16 grammes of oxygen measure 11.2 litres, then the weight will be

$$\frac{16 \times 3742}{11200} \text{ grammes; or } 5\cdot345 \text{ g.}$$

INDEX.

Acid salts, 74, 153.
Acids, theory, 150, 153.
Aethiop's Mineral, 229.
Agate, 138.
Air, a mixture, 45, 49.
Air and Nitrogen, 42, 45.
Air, Composition by Volume, 47.
    Composition by Weight, 48.
Alkali manufacture, 163.
Alkali metals, 157.
Alkalies, properties of, 160.
Alkaline Earths, 172, 173.
Alloys, 156.
Alloys, fusible, 245.
Aluminium, 188.
    salts, 189.
Alums, 190, 209.
Amalgamation, 221.
Amalgams, 225.
Ammonia, 93.
    characters, 94, 95.
    preparation, 95.
    compounds, 96.
Ammonia-Soda process, 166.
Ammonium amalgam, 170.
    nitrate, 96.
    salts, 168.
Analysis, 6.
    quantitative, 6.
    qualitative, 6.
    of gases, 32.
Antimonates, 244.
Antimonic Acids, 243.
Antimonite, 241.
Antimony, 241.
    compounds, 242.
Apatite, 112.
Aqueous Vapour in Air, 45.
Aragonite, 173.
Arsenates, 146.
Arsenic, 144.
    oxides, 145.

Arsenic acid, 146.
    hydride, 147.
    sulphides, 146.
Arsenious acid, 146.
Arsenites, 146.
Arsine, 147.
Atmospheric Air, 45.
Atomic Heat, 248.
    Theory, 7.
    Weights, 10.
    Weight and Specific Heat, 248.
Atomicity, 150.
Atoms and Molecules, 7.
Augite, 141.
Avogadro's Hypothesis, 8.

Barium, 178.
    salts, 178.
Baryta, 179.
Bases, 153.
Basicity, 153.
Bell-metal, 216.
Beryllium, 181.
Bessemer Process, 198.
Bibasic Acids, 154.
Bismuth, 245.
    salts, 246.
Bismuthic acid, 247.
Bittern, 163.
Black Ash, 163, 164.
Blast Furnace, 195.
Bleaching powder, 87.
Blowpipe flame, 61.
Boiler Crust or fur, 175.
Boiling points, 253, 255, 258.
Bone ash, 112.
Boracic acid, 136.
    as antiseptic, 137.
Borates, 137.
Borax, 137.
Boric acid, 136.

268 Index.

Boron, 135.
    allotropic forms, 135.
    compounds, 136.
    ethide, 138.
    Specific Heat, 250.
Boyle's law, 264.
Brass, 216.
Braunite, 210.
Bricks, 192.
Brimstone, 64.
Brine, 163.
Bromates, 130.
Bromine, 87.
    oxides, 125.
Bronze, 216.

Cadmium, 185.
    vapour density, 187.
    salts, 186.
Cæsium, 157.
Cairngorm, 138.
Calcite, 173.
Calcium, 173.
    salts, 174.
    phosphate, 112, 121.
Calculation of Formulæ, 155.
Calomel, 226.
    molecular weight, 228.
Carbon, 50.
    forms of, 50.
    allotropic forms, 51.
    absorptive power, 52.
Carbon di-oxide, in air, 45.
    properties, 55, 56.
    composition, 57.
    liquefaction of, 256.
Carbonic acid gas in air, 45.
Carbonic oxide, 59.
Carbon-monoxide, 59.
Carbon oxides, 53, 57, 58.
Carbon, specific heat, 250.
Carnallite, 158.
Cassiterite, 234.
Celestine, 177.
Chalcedony, 139.
Chalk, 173.
Charcoal, Animal, 51.
    Wood, 51.
Chemical Action, 3.
    Combination, Laws of, 5.
Chili Saltpetre, 162.
Chlorates, 128.
Chloric acid, 127.
Chlorides, metallic, 81.

Chlorine, 84.
    combustion in, 86.
    bleaching, 86.
    specific heat, 250.
Chlorine-oxides, 125.
Chlorous acid, 126.
    anhydride, 126.
Chromates, 207, 208.
Chrome alum, 209.
    ironstone, 206, 207.
Chromic salts, 206, 209.
Chromium, 206.
Chromous salts, 206, 209.
Cinnabar, 224.
Clarke's Soap Test, 175.
Clay, 141, 191.
Cobalt, 203.
Coinage, silver, 223.
    bronze, 216.
Coke, 51.
Colloids, 140.
Combination, 2.
Combining Weights, 7.
Combustibles, 61.
Combustion, 61.
    of carbon, 55.
    in Hydrogen, 20.
    in Oxygen, 24.
    with copper oxide, 218.
    of sugar, 218.
Compounds, 2.
Copper, 214.
    pyrites, 214.
    salts, 216.
Coprolites, 112.
Corrosive Sublimate, 227.
Corundum, 188.
Critical temperature, 259.
Cryolite, 188.
Cupellation, 236.
Cupric Salts, 217.
Cuprous Salts, 216.

Davy, Sir Humphrey, 158, 162.
Dialysis, 140, 190, 201.
Diamine, 111.
Diamond, 50.
Dibasic Acids, 74.
Diffusion, 259.
Dissociation, 169.
Distillation of Water, 40.
    of Sulphur, 67.
Distilled Water, 40.
Dulong and Petit, 249.

Index. 269

Dumas, composition of Water, 31.
Dyad elements, 150.
    Metals, 172, 181.

Earthenware, 192.
Electrolysis, 156.
Electroplating, 221.
Electrotype, 220.
Elements, 1.
Emery, 188.
Equations, 13.
Equivalent weight, 7.
Ethyl phosphamine, 123.
    phosphine, 124.
Eudiometer, described, 32.
    U-shape, 34.
    calculations, 265.
Expansion of gases, 262.

Felspar, 141, 188, 192.
Fermentation, 56.
Ferric salts, 199, 201.
Ferricyanides, 202.
Ferrocyanides, 202.
Ferrous salts, 199.
Fire clay, 192.
Flames, 62.
Flint, 139.
Flint glass, 142.
Fluoboric acid, 138.
Fluorides, 134.
Fluorine, 132.
    compounds, 134.
    isolation, 133.
Fluor spar, 173.
Formulæ, calculation of, 155.
    meaning of, 12.
Furnace, gas, 17.
    revolving, 165.
Fusible alloys, 245.

Galena, 64, 235.
Gallium, 152, 188.
Gas analysis, 32.
Gases, 255.
    diffusion, 259.
    expansion by heat, 262.
    pressure and volume, 264.
    solubility, 39.
Gas furnace, 17.
Gay Lussac Tower, 72.
German silver, 204.
Germanium, 152.

Geysers, 141.
Glass, 140, 142.
Glycerine, 161.
Graham on diffusion, 261.
Gun cotton, 99.
    metal, 216.
    powder, composition, 161.
Gypsum, 173.

Halogen elements, 78.
Hardness of water, 175.
Hard and soft waters, 40.
Hausmannite, 211.
Heat of combinations of halogens, 92, 115, 132.
Heat of combustion of carbon, 55.
    of hydrogen, 4.
    of phosphorus, 115.
Heat of formation of hydrogen chloride, 80.
Heavy spar, 180.
Heptad elements, 151.
    metals, 210.
Hexad elements, 151.
    metals, 206.
Hydriodic acid, 92.
Hydrobromic acid, 89.
Hydrochloric acid, 78, 81.
Hydro-fluo-silicic acid, 143.
Hydroxylamine, 102, 110.
Hydrogen, 14.
    atomic heat, 252.
    bromide, 88.
    chloride, 78.
    ,, composition, 82.
    fluoride, 133.
    from water, 16, 17.
    iodide, 91.
    nitrate, 98.
    properties of, 18, 19.
    sulphide, 76.
    weight and volume, 20.
Hypobromites, 130.
Hypochlorites, 127.
Hypochlorous acid, 127.
Hyponitrites, 110.
Hypophosphites, 122.
Hypophosphorous acid, 122.
Hyposulphites, 70.

Iceland spar, 173.
Indium, 188, 251.
Iodates, 130.
Iodic acid, 130.

Iodine, 89.
    molecular weight, 90.
    oxides, 125, 130.
Iron, 193.
    smelting, 194.
    cast and wrought, 194, 197.
    salts of, 200.

Jasper, 138.

Kaolin, 188.
Kelp, 89.

Latent heat, fusion, 254.
    steam, 37.
    vapours, 256.
    water, 36.
Lead and its salts, 235.
    chambers, 71.
    in glass, 142.
    tetrethyl, 231.
Leblanc, process, 163.
Lime, 176.
    milk of, 176.
    light, 17.
    soap, 175.
    water, 176.
Limestone, 173.
Litharge, 238.
Lithium, 157.
Liquefaction of gases, 256.
Liquids, 254.
Lixiviation, 158, 166.

Magnesia, 183.
Magnesium, 181.
    phosphate, 121.
Malachite, 214.
Manganates, 212.
Manganese, 210.
Manganous salts, 211.
Marble, 173.
Mariotte's Law, 264.
Marsh's Test, antimony, 245.
    arsenic, 147.
Massicot, 238.
Matter, three states, 253.
Mercuric salts, 225.
Mercurous salts, 225.
Mercury, 224.
Metals, 156.
    specific heats, 249.
Metaphosphoric acid, 119.
Mirrors, 221.

Mispickel, 144.
Mixture, 2.
Molecular Heats, 251.
Molecules, 8.
Molybdenum, 206.
Monad elements, 150.
Monobasic acids, 153.
Mortars, 176.
    hydraulic, 176.
    blue lias, 176.
Multiple proportions, 5.

Natrium, 161.
Natural waters, 39.
Neutral salts, 74.
Newlands, Mr. J. A. R., 151.
Nickel and compounds, 203.
Nitre beds, 101.
Nitric acid, 93.
    preparation, 98.
    properties, 99.
    synthesis, 100.
    with metals, 101.
Nitric anhydride, 104.
Nitrification, 100.
Nitrites, 106.
Nitrobenzene, 99.
Nitroderivatives, 99.
Nitrogen, 42, 92.
    atomic heat, 252.
    oxides, 104.
    properties, 44.
    separated from air, 42, 44.
Nitroglycerine, 99.
Nitrous acid, 106.
    oxide, 109.
Nordhausen acid, 75.

Oil of vitriol, 70, 73.
Olivine, 141.
Opal, 139.
Organic analysis, 218.
Orpiment, 144.
Oxides, reduction in Hydrogen, 30.
Oxygen, properties of, 23.
    absorbed from air, 43.
    atomic heat, 252.
    Baryta process, 180.
    from air, 21.
    in air, 49.
    liquid, 258.
    weight of, 25.

Index.

Ozone, 25.
   density of, 27.
   tube, 26.

Pattinson's process, 236.
Pearl ash, 158, 161.
Pentad elements, 151.
   metals, 241.
Pepy's gasholder, 23.
Perchlorates, 129.
Perchloric acid, 129.
Periodates, 130.
Periodic acid, 130.
   law, 152.
Permanganates, 213.
Pewter, 232.
Phosphamine, 123.
Phosphates, 120.
Phosphine, 123.
Phosphonium iodide, 123.
Phosphoric acid, 112, 118.
Phosphorite, 112.
Phosphorous acid, 121.
Phosphorus, 112.
   allotropic forms, 114.
   bromide, 88.
   chloride, 80, 116.
   heat of combustion, 115.
   iodide, 91.
   oxides, 115.
   specific heat, 250.
Photography, 224.
Plant ashes, 161, 162.
Plaster of Paris, 177.
Plating, 221.
Platinum, ammonium chloride, 169.
Porcelain, 191.
Potash, 160.
Potashes, crude, 158.
Potassium, 156, 157.
   alloy, 162.
   chlorate, 22.
   chromates, 207.
   manganate, 212.
   permanganate, 213.
   salts, 159.
Pottery, 191.
Printer's type, 237.
Prussian blue, 203.
Pyrites, 64, 194, 214.
   arsenical, 144.
Pyroboric acid, 137.
Pyrolusite, 210.
Pyrophosphoric acid, 119.

Quantivalence, 150, 151.
Quartz, 139.

Rain water, 39.
Realgar, 144.
Reciprocal proportions, 6.
Reduction, in hydrogen, 30.
Reduction of chromates, 208.
   of ferric salts, 201.
   of lead, 236.
   of oxides, 194, 217.
   of permanganates, 213.
Reinsch's test for arsenic, 148.
River water, 39.
Roasting, 236.
Rock crystal, 139.
   salt, 163.
Rubidium, 157.
Ruby, 188.

Salt cake, 163, 164.
Salterns, 163.
Salt mines, 163.
Salts, 74.
   acid, 74.
   neutral, 74.
Sand, 139.
Sapphire, 188.
Saturated solution, 38.
Schlippe's salt, 244.
Sea water, 39.
Silica, forms of, 139.
   dialysis, 140.
Silicates, 141.
Silicic acids, 140.
Silicon, 135.
   allotropic forms, 138.
   chloride, 142.
   fluoride, 142.
   hydride, 143.
   oxide, 139.
   specific heat, 250.
Silver, 220.
   coinage, 223.
   from lead, 236.
   salts, 223.
Silvering glass, 221.
Silvine, 158.
Sinter, 141.
Smalt, 204.
Soap, lime, 175.
   soft, 161.
Soaps, production of, 161.

Soda, 167.
  ash, 163.
  crystals, 166.
Sodium, 156, 161.
  alloy, 162.
  borate, 137.
  phosphate, 120.
  salts, 162, 166.
Softening waters, 175.
Solder, 232.
Solution, 37, 38.
Soot, 50, 51.
Specific heats, 249, 250.
Spinel, 188.
Spring water, 39.
Stannates, 234.
Stannic acids, 234.
Stannic salts, 232.
Stannous salts, 232.
Stannum, Tin, 231.
Stassfurt, salt mines, 158.
Steam, decomposed, 17, 28.
  latent heat of, 37, 256.
  vapour density, 35.
Steel, 198.
Stibine, 244.
Stibium, 241.
Stoneware, 192.
Strontia, 178.
Strontianite, 177.
Strontium, 177.
  salts, 178.
Structure of flame, 61.
Sulph-antimonates, 244.
Sulph-antimonites, 244.
Sulpharsenates, 146.
Sulpharsenites, 146.
Sulphites, 69.
Sulpho-stannates, 235.
Sulphur, 64.
  allotropic forms, 65.
  dioxide, 68.
  oxides, 68.
  specific heat, 250.
  trioxide, 74.
  vapour density, 67.
  varieties, 64.
Sulphuretted hydrogen, 76.
Sulphuric anhydride, 74.
Sulphurous acid, 69.
Symbols, use of, 11.

Tables—
  Atomic weights, 11.
  Periodic law, 152.
  Specific heats, 249, 250.
Tartar Emetic, 243.
Tetrad elements, 151.
  metals, 231.
Thallium, 188.
Thermal Unit, 4, 36, 37.
Thiosulphates, 70.
Triad elements, 150.
  metals, 188.
Tribasic acids, 154.
Tin, 231.
  alloys, 232.
  plate, 232.
  salts, 232.
  stone, 231.
  tetrethyl, 231.
Tincal, 135.
Tungsten, 206.
Turnbull's Blue, 203.
Type-metal, 237.

Uranium, 206.

Valency, 150.
Vapours, 253, 255.
Ventilation, 57.
Vermilion, 229.

Water at 4° C, 36.
  boiling, 35.
   ,,   under reduced pressure, 35.
  composition by volume, 32.
  composition by weight, 30, 31.
  condensation as rain, etc., 45.
  decomposition of, 16, 17, 28, 29.
  in air, 45.
  latent heat of, 36, 254.
  solvent action, 37.
  synthesis, 29, 30.
  vapour density, 35.
Well water, 39.
White lead, 239.
Witherite, 180.

Zinc, 184.
  ethyl, 187.
  salts, 185.

THE END.

July, 1888.

Clarendon Press, Oxford.

A SELECTION OF

BOOKS

PUBLISHED FOR THE UNIVERSITY BY

HENRY FROWDE,

AT THE OXFORD UNIVERSITY PRESS WAREHOUSE,
AMEN CORNER, LONDON.

ALSO TO BE HAD AT THE
CLARENDON PRESS DEPOSITORY, OXFORD.

[*Every book is bound in cloth, unless otherwise described.*]

LEXICONS, GRAMMARS, ORIENTAL WORKS, &c.

ANGLO-SAXON.—*An Anglo-Saxon Dictionary*, based on the MS. Collections of the late Joseph Bosworth, D.D., Professor of Anglo-Saxon, Oxford. Edited and enlarged by Prof. T. N. Toller, M.A. (To be completed in four parts.) Parts I-III. A—SAR. 4to. 15*s.* each.

ARABIC.—*A Practical Arabic Grammar.* Part I. Compiled by A. O. Green, Brigade Major, Royal Engineers, Author of 'Modern Arabic Stories.' Second Edition, Enlarged and Revised. Crown 8vo. 7*s.* 6*d.*

CHINESE.—*A Handbook of the Chinese Language.* By James Summers. 1863. 8vo. half bound, 1*l.* 8*s.*

—— *A Record of Buddhistic Kingdoms*, by the Chinese Monk FÂ-HIEN. Translated and annotated by James Legge, M.A., LL.D. Crown 4to. cloth back, 10*s.* 6*d.*

ENGLISH.—*A New English Dictionary, on Historical Principles:* founded mainly on the materials collected by the Philological Society. Edited by James A. H. Murray, LL.D., with the assistance of many Scholars and men of Science. Vol. I. A and B. Imperial 4to. half Morocco, 2*l.* 12*s.* 6*d.*

Part IV, Section II, C—CASS. Beginning of Vol. II, 5*s.*

B

ENGLISH.—*An Etymological Dictionary of the English Language*. By W. W. Skeat, Litt.D. Second Edition. 1884. 4to. 2*l*. 4*s*.

——Supplement to the First Edition of the above. 4to. 2*s*. 6*d*.

—— *A Concise Etymological Dictionary of the English Language*. By W. W. Skeat, Litt.D. Third Edition. 1887. Crown 8vo. 5*s*. 6*d*.

GREEK.—*A Greek-English Lexicon*, by Henry George Liddell, D.D., and Robert Scott, D.D. Seventh Edition, Revised and Augmented throughout. 1883. 4to. 1*l*. 16*s*.

—— *A Greek-English Lexicon*, abridged from Liddell and Scott's 4to. edition, chiefly for the use of Schools. Twenty-first Edition. 1884. Square 12mo. 7*s*. 6*d*.

—— *A copious Greek-English Vocabulary*, compiled from the best authorities. 1850. 24mo. 3*s*.

—— *A Practical Introduction to Greek Accentuation*, by H. W. Chandler, M.A. Second Edition. 1881. 8vo. 10*s*. 6*d*.

HEBREW.—*The Book of Hebrew Roots*, by Abu 'l-Walîd Marwân ibn Janâh, otherwise called Rabbî Yônâh. Now first edited, with an Appendix, by Ad. Neubauer. 1875. 4to. 2*l*. 7*s*. 6*d*.

—— *A Treatise on the use of the Tenses in Hebrew*. By S. R. Driver, D.D. Second Edition. 1881. Extra fcap. 8vo. 7*s*. 6*d*.

—— *Hebrew Accentuation of Psalms, Proverbs, and Job*. By William Wickes, D.D. 1881. Demy 8vo. 5*s*.

—— *A Treatise on the Accentuation of the twenty-one so-called Prose Books of the Old Testament*. By William Wickes, D.D. 1887. Demy 8vo. 10*s*. 6*d*.

ICELANDIC.—*An Icelandic-English Dictionary*, based on the MS. collections of the late Richard Cleasby. Enlarged and completed by G. Vigfússon, M.A. With an Introduction, and Life of Richard Cleasby, by G. Webbe Dasent, D.C.L. 1874. 4to. 3*l*. 7*s*.

—— *A List of English Words the Etymology of which is illustrated by comparison with Icelandic*. Prepared in the form of an APPENDIX to the above. By W. W. Skeat, Litt.D. 1876. stitched, 2*s*.

—— *An Icelandic Primer*, with Grammar, Notes, and Glossary. By Henry Sweet, M.A. Extra fcap. 8vo. 3*s*. 6*d*.

—— *An Icelandic Prose Reader*, with Notes, Grammar and Glossary, by Dr. Gudbrand Vigfússon and F. York Powell, M.A. 1879. Extra fcap. 8vo. 10*s*. 6*d*.

LATIN.—*A Latin Dictionary*, founded on Andrews' edition of Freund's Latin Dictionary, revised, enlarged, and in great part rewritten by Charlton T. Lewis, Ph.D., and Charles Short, LL.D. 1879. 4to. 1*l*. 5*s*.

MELANESIAN.—*The Melanesian Languages.* By R. H. Codrington, D.D., of the Melanesian Mission. 8vo. 18s.

SANSKRIT.—*A Practical Grammar of the Sanskrit Language*, arranged with reference to the Classical Languages of Europe, for the use of English Students, by Sir M. Monier-Williams, M.A. Fourth Edition. 8vo. 15s.

—— *A Sanskrit-English Dictionary*, Etymologically and Philologically arranged, with special reference to Greek, Latin, German, Anglo-Saxon, English, and other cognate Indo-European Languages. By Sir M. Monier-Williams, M.A. 1888. 4to. 4l. 14s. 6d.

—— *Nalopákhyánam.* Story of Nala, an Episode of the Mahá-Bhárata: the Sanskrit text, with a copious Vocabulary, and an improved version of Dean Milman's Translation, by Sir M. Monier-Williams, M.A. Second Edition, Revised and Improved. 1879. 8vo. 15s.

—— *Sakuntalā.* A Sanskrit Drama, in Seven Acts. Edited by Sir M. Monier-Williams, M.A. Second Edition, 1876. 8vo. 21s.

SYRIAC.—*Thesaurus Syriacus*: collegerunt Quatremère, Bernstein, Lorsbach, Arnoldi, Agrell, Field, Roediger: edidit R. Payne Smith, S.T.P. Fasc. I-VI. 1868-83. sm. fol. each, 1l. 1s. Fasc. VII. 1l. 11s. 6d. Vol. I, containing Fasc. I-V, sm. fol. 5l. 5s.

—— *The Book of Kalīlah and Dimnah.* Translated from Arabic into Syriac. Edited by W. Wright, LL.D. 1884. 8vo. 21s.

GREEK CLASSICS, &c.

Aristophanes: A Complete Concordance to the Comedies and Fragments. By Henry Dunbar, M.D. 4to. 1l. 1s.

Aristotle: The Politics, with Introductions, Notes, etc., by W. L. Newman, M.A., Fellow of Balliol College, Oxford. Vols. I. and II. Medium 8vo. 28s.

Aristotle: The Politics, translated into English, with Introduction, Marginal Analysis, Notes, and Indices, by B. Jowett, M.A. Medium 8vo. 2 vols. 21s.

Catalogus Codicum Graecorum Sinaiticorum. Scripsit V. Gardthausen Lipsiensis. With six pages of Facsimiles. 8vo. *linen*, 25s.

Heracliti Ephesii Reliquiae. Recensuit I. Bywater, M.A. Appendicis loco additae sunt Diogenis Laertii Vita Heracliti, Particulae Hippocratei De Diaeta Libri Primi, Epistolae Heracliteae. 1877. 8vo. 6s.

Herculanensium Voluminum Partes II. 1824. 8vo. 10s.

Fragmenta Herculanensia. A Descriptive Catalogue of the Oxford copies of the Herculanean Rolls, together with the texts of several papyri, accompanied by facsimiles. Edited by Walter Scott, M.A., Fellow of Merton College, Oxford. Royal 8vo. *cloth*, 21*s*.

Homer: A Complete Concordance to the Odyssey and Hymns of Homer; to which is added a Concordance to the Parallel Passages in the Iliad, Odyssey, and Hymns. By Henry Dunbar, M.D. 1880. 4to. 1*l*. 1*s*.

—— *Scholia Graeca in Iliadem.* Edited by Professor W. Dindorf, after a new collation of the Venetian MSS. by D. B. Monro, M.A., Provost of Oriel College. 4 vols. 8vo. 2*l*. 10*s*. Vols. V and VI. *In the Press.*

—— *Scholia Graeca in Odysseam.* Edidit Guil. Dindorfius. Tomi II. 1855. 8vo. 15*s*. 6*d*.

Plato: Apology, with a revised Text and English Notes, and a Digest of Platonic Idioms, by James Riddell, M.A. 1878. 8vo. 8*s*. 6*d*.

—— *Philebus,* with a revised Text and English Notes, by Edward Poste, M.A. 1860. 8vo. 7*s*. 6*d*.

—— *Sophistes and Politicus,* with a revised Text and English Notes, by L. Campbell, M.A. 1867. 8vo. 18*s*.

—— *Theaetetus,* with a revised Text and English Notes, by L. Campbell, M.A. Second Edition. 8vo. 10*s*. 6*d*.

—— *The Dialogues,* translated into English, with Analyses and Introductions, by B. Jowett, M.A. A new Edition in 5 volumes, medium 8vo. 1875. 3*l*. 10*s*.

—— *The Republic,* translated into English, with an Analysis and Introduction, by B. Jowett, M.A. Medium 8vo. 12*s*. 6*d*.

Thucydides: Translated into English, with Introduction, Marginal Analysis, Notes, and Indices. By B. Jowett, M.A. 2 vols. 1881. Medium 8vo. 1*l*. 12*s*.

THE HOLY SCRIPTURES, &c.

STUDIA BIBLICA.—Essays in Biblical Archæology and Criticism, and kindred subjects. By Members of the University of Oxford. 8vo. 10*s*. 6*d*.

ENGLISH.—*The Holy Bible in the earliest English Versions,* made from the Latin Vulgate by John Wycliffe and his followers: edited by the Rev. J. Forshall and Sir F. Madden. 4 vols. 1850. Royal 4to. 3*l*. 3*s*.

[Also reprinted from the above, with Introduction and Glossary by W. W. Skeat, Litt. D.

ENGLISH.—*The Books of Job, Psalms, Proverbs, Ecclesiastes, and the Song of Solomon:* according to the Wycliffite Version made by Nicholas de Hereford, about A.D. 1381, and Revised by John Purvey, about A.D. 1388. Extra fcap. 8vo. 3*s.* 6*d.*

—— *The New Testament in English,* according to the Version by John Wycliffe, about A.D. 1380, and Revised by John Purvey, about A.D. 1388. Extra fcap. 8vo. 6*s.*]

—— *The Holy Bible:* an exact reprint, page for page, of the Authorised Version published in the year 1611. Demy 4to. half bound, 1*l.* 1*s.*

—— *The Psalter, or Psalms of David, and certain Canticles,* with a Translation and Exposition in English, by Richard Rolle of Hampole. Edited by H. R. Bramley, M.A., Fellow of S. M. Magdalen College, Oxford. With an Introduction and Glossary. Demy 8vo. 1*l.* 1*s.*

—— *Lectures on the Book of Job.* Delivered in Westminster Abbey by the Very Rev. George Granville Bradley, D.D., Dean of Westminster. Crown 8vo. 7*s.* 6*d.*

—— *Lectures on Ecclesiastes.* By the same Author. Crown 8vo. 4*s.* 6*d.*

GOTHIC.—*The Gospel of St. Mark in Gothic,* according to the translation made by Wulfila in the Fourth Century. Edited with a Grammatical Introduction and Glossarial Index by W. W. Skeat, Litt. D. Extra fcap. 8vo. 4*s.*

GREEK.—*Vetus Testamentum* ex Versione Septuaginta Interpretum secundum exemplar Vaticanum Romae editum. Accedit potior varietas Codicis Alexandrini. Tomi III. Editio Altera. 18mo. 18*s.* The volumes may be had separately, price 6*s.* each.

—— *Origenis Hexaplorum* quae supersunt; sive, Veterum Interpretum Graecorum in totum Vetus Testamentum Fragmenta. Edidit Fridericus Field, A.M. 2 vols. 1875. 4to. 5*l.* 5*s.*

—— *The Book of Wisdom:* the Greek Text, the Latin Vulgate, and the Authorised English Version; with an Introduction, Critical Apparatus, and a Commentary. By William J. Deane, M.A. Small 4to. 12*s.* 6*d.*

—— *Novum Testamentum Graece.* Antiquissimorum Codicum Textus in ordine parallelo dispositi. Accedit collatio Codicis Sinaitici. Edidit E. H. Hansell, S. T. B. Tomi III. 1864. 8vo. 24*s.*

—— *Novum Testamentum Graece.* Accedunt parallela S. Scripturae loca, etc. Edidit Carolus Lloyd, S.T.P.R. 18mo. 3*s.*

On writing paper, with wide margin, 10*s.*

GREEK.—*Novum Testamentum Graece* juxta Exemplar Millianum. 18mo. 2s. 6d. On writing paper, with wide margin, 9s.

—— *Evangelia Sacra Graece.* Fcap. 8vo. limp, 1s. 6d.

—— *The Greek Testament*, with the Readings adopted by the Revisers of the Authorised Version:—

(1) Pica type, with Marginal References. Demy 8vo. 10s. 6d.
(2) Long Primer type. Fcap. 8vo. 4s. 6d.
(3) The same, on writing paper, with wide margin, 15s.

—— *The Parallel New Testament*, Greek and English; being the Authorised Version, 1611; the Revised Version, 1881; and the Greek Text followed in the Revised Version. 8vo. 12s. 6d.

The Revised Version is the joint property of the Universities of Oxford and Cambridge.

—— *Canon Muratorianus:* the earliest Catalogue of the Books of the New Testament. Edited with Notes and a Facsimile of the MS. in the Ambrosian Library at Milan, by S. P. Tregelles, LL.D. 1867. 4to. 10s. 6d.

—— *Outlines of Textual Criticism applied to the New Testament.* By C. E. Hammond, M.A. Fourth Edition. Extra fcap. 8vo. 3s. 6d.

HEBREW, etc.—*Notes on the Hebrew Text of the Book of Genesis.* With Two Appendices. By G. J. Spurrell, M.A. Crown 8vo. 10s. 6d.

—— *The Psalms in Hebrew without points.* 1879. Crown 8vo. Price reduced to 2s., in stiff cover.

—— *A Commentary on the Book of Proverbs.* Attributed to Abraham Ibn Ezra. Edited from a MS. in the Bodleian Library by S. R. Driver, M.A. Crown 8vo. paper covers, 3s. 6d.

—— *The Book of Tobit.* A Chaldee Text, from a unique MS. in the Bodleian Library; with other Rabbinical Texts, English Translations, and the Itala. Edited by Ad. Neubauer, M.A. 1878. Crown 8vo. 6s.

—— *Horae Hebraicae et Talmudicae*, a J. Lightfoot. A new Edition, by R. Gandell, M.A. 4 vols. 1859. 8vo. 1l. 1s.

LATIN.—*Libri Psalmorum* Versio antiqua Latina, cum Paraphrasi Anglo-Saxonica. Edidit B. Thorpe, F.A.S. 1835. 8vo. 10s. 6d.

—— *Old-Latin Biblical Texts: No. I.* The Gospel according to St. Matthew from the St. Germain MS. (g_1). Edited with Introduction and Appendices by John Wordsworth, D.D. Small 4to., stiff covers, 6s.

—— *Old-Latin Biblical Texts: No. II.* Portions of the Gospels according to St. Mark and St. Matthew, from the Bobbio MS. (k), &c. Edited by John Wordsworth, D.D., W. Sanday, M.A., D.D., and H. J. White, M.A. Small 4to., stiff covers, 21s.

LATIN.—*Old-Latin Biblical Texts: No. III.* The Four Gospels, from the Munich MS. (q), now numbered Lat. 6224 in the Royal Library at Munich. With a Fragment from St. John in the Hof-Bibliothek at Vienna (Cod. Lat. 502). Edited, with the aid of Tischendorf's transcript (under the direction of the Bishop of Salisbury), by H. J. White, M.A. Small 4to. stiff covers, 12s. 6d.

OLD-FRENCH.—*Libri Psalmorum* Versio antiqua Gallica e Cod. MS. in Bibl. Bodleiana adservato, una cum Versione Metrica aliisque Monumentis pervetustis. Nunc primum descripsit et edidit Franciscus Michel, Phil. Doc. 1860. 8vo. 10s. 6d.

FATHERS OF THE CHURCH, &c.

St. Athanasius: Historical Writings, according to the Benedictine Text. With an Introduction by William Bright, D.D. 1881. Crown 8vo. 10s. 6d.

—— *Orations against the Arians.* With an Account of his Life by William Bright, D.D. 1873. Crown 8vo. 9s.

St. Augustine: Select Anti-Pelagian Treatises, and the Acts of the Second Council of Orange. With an Introduction by William Bright, D.D. Crown 8vo. 9s.

Canons of the First Four General Councils of Nicaea, Constantinople, Ephesus, and Chalcedon. 1877. Crown 8vo. 2s. 6d.

—— *Notes on the Canons of the First Four General Councils.* By William Bright, D.D. 1882. Crown 8vo. 5s. 6d.

Cyrilli Archiepiscopi Alexandrini in XII Prophetas. Edidit P. E. Pusey, A.M. Tomi II. 1868. 8vo. cloth, 2l. 2s.

—— *in D. Joannis Evangelium.* Accedunt Fragmenta varia necnon Tractatus ad Tiberium Diaconum duo. Edidit post Aubertum P. E. Pusey, A.M. Tomi III. 1872. 8vo. 2l. 5s.

—— *Commentarii in Lucae Evangelium* quae supersunt Syriace. E MSS. apud Mus. Britan. edidit R. Payne Smith, A.M. 1858. 4to. 1l. 2s.

—— Translated by R. Payne Smith, M.A. 2 vols. 1859. 8vo. 14s.

Ephraemi Syri, Rabulae Episcopi Edesseni, Balaei, aliorumque Opera Selecta. E Codd. Syriacis MSS. in Museo Britannico et Bibliotheca Bodleiana asservatis primus edidit J. J. Overbeck. 1865. 8vo. 1l. 1s.

Eusebius' Ecclesiastical History, according to the text of Burton, with an Introduction by William Bright, D.D. 1881. Crown 8vo. 8s. 6d.

Irenaeus: The Third Book of St. Irenaeus, Bishop of Lyons, against Heresies. With short Notes and a Glossary by H. Deane, B.D. 1874. Crown 8vo. 5s. 6d.

Patrum Apostolicorum, S. Clementis Romani, S. Ignatii, S. Polycarpi, quae supersunt. Edidit Guil. Jacobson, S.T.P.R. Tomi II. Fourth Edition, 1863. 8vo. 1l. 1s.

Socrates' Ecclesiastical History, according to the Text of Hussey, with an Introduction by William Bright, D.D. 1878. Crown 8vo. 7s. 6d.

ECCLESIASTICAL HISTORY, BIOGRAPHY, &c.

Ancient Liturgy of the Church of England, according to the uses of Sarum, York, Hereford, and Bangor, and the Roman Liturgy arranged in parallel columns, with preface and notes. By William Maskell, M.A. Third Edition. 1882. 8vo. 15s.

Baedae Historia Ecclesiastica. Edited, with English Notes, by G. H. Moberly, M.A. 1881. Crown 8vo. 10s. 6d.

Bright (W.). Chapters of Early English Church History. 1878. 8vo. 12s.

Burnet's History of the Reformation of the Church of England. A new Edition. Carefully revised, and the Records collated with the originals, by N. Pocock, M.A. 7 vols. 1865. 8vo. *Price reduced to* 1l. 10s.

Councils and Ecclesiastical Documents relating to Great Britain and Ireland. Edited, after Spelman and Wilkins, by A. W. Haddan, B.D., and W. Stubbs, M.A. Vols. I. and III. 1869-71. Medium 8vo. each 1l. 1s.

    Vol. II. Part I. 1873. Medium 8vo. 10s. 6d.

    Vol. II. Part II. 1878. Church of Ireland; Memorials of St. Patrick. Stiff covers, 3s. 6d.

Hamilton (John, Archbishop of St. Andrews), The Catechism of. Edited, with Introduction and Glossary, by Thomas Graves Law. With a Preface by the Right Hon. W. E. Gladstone. 8vo. 12s. 6d.

Hammond (C. E.). Liturgies, Eastern and Western. Edited, with Introduction, Notes, and Liturgical Glossary. 1878. Crown 8vo. 10s. 6d.

    An Appendix to the above. 1879. Crown 8vo. paper covers, 1s. 6d.

John, Bishop of Ephesus. The Third Part of his Ecclesiastical History. [In Syriac.] Now first edited by William Cureton, M.A. 1853. 4to. 1l. 12s.

—— Translated by R. Payne Smith, M.A. 1860. 8vo. 10s.

Leofric Missal, The, as used in the Cathedral of Exeter during the Episcopate of its first Bishop, A.D. 1050-1072; together with some Account of the Red Book of Derby, the Missal of Robert of Jumièges, and a few other early MS. Service Books of the English Church. Edited, with Introduction and Notes, by F. E. Warren, B.D. 4to. half morocco, 35*s.*

Monumenta Ritualia Ecclesiae Anglicanae. The occasional Offices of the Church of England according to the old use of Salisbury, the Prymer in English, and other prayers and forms, with dissertations and notes. By William Maskell, M.A. Second Edition. 1882. 3 vols. 8vo. 2*l.* 10*s.*

Records of the Reformation. The Divorce, 1527-1533. Mostly now for the first time printed from MSS. in the British Museum and other libraries. Collected and arranged by N. Pocock, M.A. 1870. 2 vols. 8vo. 1*l.* 16*s.*

Shirley (W. W.). Some Account of the Church in the Apostolic Age. Second Edition, 1874. Fcap. 8vo. 3*s.* 6*d.*

Stubbs (W.). Registrum Sacrum Anglicanum. An attempt to exhibit the course of Episcopal Succession in England. 1858. Small 4to. 8*s.* 6*d.*

Warren (F. E.). Liturgy and Ritual of the Celtic Church. 1881. 8vo. 14*s.*

ENGLISH THEOLOGY.

Bampton Lectures, 1886. *The Christian Platonists of Alexandria.* By Charles Bigg, D.D. 8vo. 10*s.* 6*d.*

Butler's Works, with an Index to the Analogy. 2 vols. 1874. 8vo. 11*s.*

Also separately,

Sermons, 5*s.* 6*d.*   *Analogy of Religion,* 5*s.* 6*d.*

Greswell's Harmonia Evangelica. Fifth Edition. 8vo. 9*s.* 6*d.*

Heurtley's Harmonia Symbolica: Creeds of the Western Church. 1858. 8vo. 6*s.* 6*d.*

Homilies appointed to be read in Churches. Edited by J. Griffiths, M.A. 1859. 8vo. 7*s.* 6*d.*

Hooker's Works, with his life by Walton, arranged by John Keble, M.A. Seventh Edition. *Revised by R. W. Church, M.A., D.C.L., Dean of St. Paul's, and F. Paget, D.D.* 3 vols. medium 8vo. 36*s.*

Hooker's Works, the text as arranged by John Keble, M.A. 2 vols. 1875. 8vo. 11*s.*

Jewel's Works. Edited by R. W. Jelf, D.D. 8 vols. 1848. 8vo. 1*l.* 10*s.*

Pearson's Exposition of the Creed. Revised and corrected by E. Burton, D.D. Sixth Edition, 1877. 8vo. 10s. 6d.

Waterland's Review of the Doctrine of the Eucharist, with a Preface by the late Bishop of London. Crown 8vo. 6s. 6d.

—— *Works*, with Life, by Bp. Van Mildert. A new Edition, with copious Indexes. 6 vols. 1856. 8vo. 2l. 11s.

Wheatly's Illustration of the Book of Common Prayer. A new Edition, 1846. 8vo. 5s.

Wyclif. A Catalogue of the Original Works of John Wyclif, by W. W. Shirley, D.D. 1865. 8vo. 3s. 6d.

—— *Select English Works.* By T. Arnold, M.A. 3 vols. 1869-1871. 8vo. 1l. 1s.

—— *Trialogus.* With the Supplement now first edited. By Gotthard Lechler. 1869. 8vo. 7s.

HISTORICAL AND DOCUMENTARY WORKS.

British Barrows, a Record of the Examination of Sepulchral Mounds in various parts of England. By William Greenwell, M.A., F.S.A. Together with Description of Figures of Skulls, General Remarks on Prehistoric Crania, and an Appendix by George Rolleston, M.D., F.R.S. 1877. Medium 8vo. 25s.

Clarendon's History of the Rebellion and Civil Wars in England. Re-edited from a fresh Collation of the Original MS. in the Bodleian Library, with Marginal Dates, and Occasional Notes, by W. Dunn Macray, M.A., F.S.A. In six volumes, crown 8vo. *cloth*, 2l. 5s.

Clarendon's History of the Rebellion and Civil Wars in England. Also his Life, written by himself, in which is included a Continuation of his History of the Grand Rebellion. With copious Indexes. In one volume, royal 8vo. 1842. 1l. 2s.

Clinton's Epitome of the Fasti Hellenici. 1851. 8vo. 6s. 6d.

—— *Epitome of the Fasti Romani.* 1854. 8vo. 7s.

Corpvs Poeticvm Boreale. The Poetry of the Old Northern Tongue, from the Earliest Times to the Thirteenth Century. Edited, classified, and translated, with Introduction, Excursus, and Notes, by Gudbrand Vigfússon, M.A., and F. York Powell, M.A. 2 vols. 1883. 8vo. 42s.

Earle (J., M.A.). A Handbook to the Land-Charters, and other Saxonic Documents. Crown 8vo. *cloth*, 16s.

Freeman (E. A.). History of the Norman Conquest of England; its Causes and Results. In Six Volumes. 8vo. 5*l.* 9*s.* 6*d.*

—— *The Reign of William Rufus and the Accession of* Henry the First. 2 vols. 8vo. 1*l.* 16*s.*

Gascoigne's Theological Dictionary ("Liber Veritatum"): Selected Passages, illustrating the condition of Church and State, 1403-1458. With an Introduction by James E. Thorold Rogers, M.A. Small 4to. 10*s.* 6*d.*

Johnson (Samuel, LL.D.), Boswell's Life of; including Boswell's Journal of a Tour to the Hebrides, and Johnson's Diary of a Journey into North Wales. Edited by G. Birkbeck Hill, D.C.L. In six volumes, medium 8vo. With Portraits and Facsimiles of Handwriting. Half bound, 3*l.* 3*s.* (See p. 21.)

Magna Carta, a careful Reprint. Edited by W. Stubbs, D.D. 1879. 4to. stitched, 1*s.*

Passio et Miracula Beati Olaui. Edited from a Twelfth-Century MS. in the Library of Corpus Christi College, Oxford, by Frederick Metcalfe, M.A. Small 4to. stiff covers, 6*s.*

Protests of the Lords, including those which have been expunged, from 1624 to 1874; with Historical Introductions. Edited by James E. Thorold Rogers, M.A. 1875. 3 vols. 8vo. 2*l.* 2*s.*

Rogers (J. E. T.). History of Agriculture and Prices in England, A.D. 1259-1793.
Vols. I—VI (1259-1702). 8vo. 7*l.* 2*s.*

—— *The First Nine Years of the Bank of England.* 8vo. 8*s.* 6*d.*

Stubbs (W., D.D.). Seventeen Lectures on the Study of Medieval and Modern History, &c., delivered at Oxford 1867-1884. Crown 8vo. 8*s.* 6*d.*

Sturlunga Saga, including the Islendinga Saga of Lawman Sturla Thordsson and other works. Edited by Dr. Gudbrand Vigfússon. In 2 vols. 1878. 8vo. 2*l.* 2*s.*

York Plays. The Plays performed by the Crafts or Mysteries of York on the day of Corpus Christi in the 14th, 15th, and 16th centuries. Now first printed from the unique MS. in the Library of Lord Ashburnham. Edited with Introduction and Glossary by Lucy Toulmin Smith. 8vo. 21*s.*

Manuscript Materials relating to the History of Oxford. Arranged by F. Madan, M.A. 8vo. 7*s.* 6*d.*

Statutes of the University of Oxford, codified in the year 1636 under the authority of Archbishop Laud. Edited by the late J. Griffiths, D.D., with an Introduction on the History of the Laudian Code, by C. L. Shadwell, M.A., B.C.L. 4to. 1*l.* 1*s.*

Statutes made for the University of Oxford, and for the Colleges and Halls therein, by the University of Oxford Commissioners. 1882. 8vo. 12s. 6d.

Statutes supplementary to the above, approved by the Queen in Council, 1882–1888. 8vo. 2s. 6d.

Statuta Universitatis Oxoniensis. 1887. 8vo. 5s.

The Oxford University Calendar for the year 1888. Crown 8vo. 4s. 6d.

The present Edition includes all Class Lists and other University distinctions for the eight years ending with 1887.

Also, supplementary to the above, price 5s. (pp. 606),

The Honours Register of the University of Oxford. A complete Record of University Honours, Officers, Distinctions, and Class Lists; of the Heads of Colleges, &c., &c., from the Thirteenth Century to 1883.

The Examination Statutes for the Degrees of B.A., B. Mus., B.C.L., and B.M. Revised to the end of Michaelmas Term, 1887. 8vo. sewed, 1s.

The Student's Handbook to the University and Colleges of Oxford. Ninth Edition. Crown 8vo. 2s. 6d.

MATHEMATICS, PHYSICAL SCIENCE, &c.

Acland (H. W., M.D., F.R.S.). Synopsis of the Pathological Series in the Oxford Museum. 1867. 8vo. 2s. 6d.

Annals of Botany. Edited by Isaac Bayley Balfour, M.A., M.D., F.R.S., Sydney H. Vines, D.Sc., F.R.S., and William Gilson Farlow, M.D., Professor of Cryptogamic Botany in Harvard University, Cambridge, Mass., U.S.A., and other Botanists. Royal 8vo. Vol. I., half morocco, 1l. 16s. Vol. II. No. 1. *Just Published.*

Burdon-Sanderson (J., M.D., F.R.SS. L. and E.). Translations of Foreign Biological Memoirs. I. Memoirs on the Physiology of Nerve, of Muscle, and of the Electrical Organ. Medium 8vo. 21s.

De Bary (Dr. A.). Comparative Anatomy of the Vegetative Organs of the Phanerogams and Ferns. Translated and Annotated by F. O. Bower, M.A., F.L.S., and D. H. Scott, M.A., Ph.D., F.L.S. With 241 woodcuts and an Index. Royal 8vo., half morocco, 1l. 2s. 6d.

*De Bary (Dr. A.) Comparative Morphology and Biology of the
Fungi Mycetozoa and Bacteria*. Authorised English Translation by Henry
E. F. Garnsey, M.A. Revised by Isaac Bayley Balfour, M.A., M.D., F.R.S.
With 198 Woodcuts. Royal 8vo., half morocco, 1*l*. 2*s*. 6*d*

—— *Lectures on Bacteria*. Second improved edition. Authorised translation by H. E. F. Garnsey, M.A. Revised by Isaac Bayley Balfour, M.A., M.D., F.R.S. With 20 Woodcuts. Crown 8vo. 6*s*.

Goebel (Dr. K.). Outlines of Classification and Special Morphology of Plants. A New Edition of Sachs' Text-Book of Botany, Book II. English Translation by H. E. F. Garnsey, M.A. Revised by I. Bayley Balfour, M.A., M.D., F.R.S. With 407 Woodcuts. Royal 8vo. half morocco, 21*s*.

Müller (J.). On certain Variations in the Vocal Organs of the Passeres that have hitherto escaped notice. Translated by F. J. Bell, B.A., and edited, with an Appendix, by A. H. Garrod, M.A., F.R.S. With Plates. 1878. 4to. paper covers, 7*s*. 6*d*.

Price (Bartholomew, M.A., F.R.S.). Treatise on Infinitesimal Calculus.

Vol. I. Differential Calculus. Second Edition. 8vo. 14*s*. 6*d*.

Vol. II. Integral Calculus, Calculus of Variations, and Differential Equations. Second Edition, 1865. 8vo. 18*s*.

Vol. III. Statics, including Attractions; Dynamics of a Material Particle. Second Edition, 1868. 8vo. 16*s*.

Vol. IV. Dynamics of Material Systems; together with a chapter on Theoretical Dynamics, by W. F. Donkin, M.A., F.R.S. 1862. 8vo. 16*s*.

Pritchard (C., D.D., F.R.S.). Uranometria Nova Oxoniensis.
A Photometric determination of the magnitudes of all Stars visible to the naked eye, from the Pole to ten degrees south of the Equator. 1885. Royal 8vo. 8*s*. 6*d*.

—— *Astronomical Observations* made at the University Observatory, Oxford, under the direction of C. Pritchard, D.D. No. 1. 1878. Royal 8vo. paper covers, 3*s*. 6*d*.

Rigaud's Correspondence of Scientific Men of the 17th Century, with Table of Contents by A. de Morgan, and Index by the Rev. J. Rigaud, M.A. 2 vols. 1841-1862. 8vo. 18*s*. 6*d*.

Rolleston (George, M.D., F.R.S.). Forms of Animal Life.
A Manual of Comparative Anatomy, with descriptions of selected types. Second Edition. Revised and enlarged by W. Hatchett Jackson, M.A. Medium, 8vo. cloth extra, 1*l*. 16*s*.

—— *Scientific Papers and Addresses*. Arranged and Edited by William Turner, M.B., F.R.S. With a Biographical Sketch by Edward Tylor, F.R.S. With Portrait, Plates, and Woodcuts. 2 vols. 8vo. 1*l*. 4*s*.

Sachs (Julius von). Lectures on the Physiology of Plants.
Translated by H. Marshall Ward, M.A. With 445 Woodcuts. Royal 8vo. half morocco, 1*l*. 11*s*. 6*d*.

Westwood (J. O., M.A., F.R.S.). Thesaurus Entomologicus Hopeianus, or a Description of the rarest Insects in the Collection given to the University by the Rev. William Hope. With 40 Plates. 1874. Small folio, half morocco, 7*l.* 10*s.*

The Sacred Books of the East.

TRANSLATED BY VARIOUS ORIENTAL SCHOLARS, AND EDITED BY
F. MAX MÜLLER.

[Demy 8vo. cloth.]

Vol. I. The Upanishads. Translated by F. Max Müller. Part I. The *Kh*ândogya-upanishad, The Talavakâra-upanishad, The Aitareya-âra*n*yaka, The Kaushîtaki-brâhma*n*a-upanishad, and The Vâ*g*asaneyi-sa*m*hitâ-upanishad. 10*s.* 6*d.*

Vol. II. The Sacred Laws of the Âryas, as taught in the Schools of Âpastamba, Gautama, Vâsish*th*a, and Baudhâyana. Translated by Prof. Georg Bühler. Part I. Âpastamba and Gautama. 10*s.* 6*d.*

Vol. III. The Sacred Books of China. The Texts of Confucianism. Translated by James Legge. Part I. The Shû King, The Religious portions of the Shih King, and The Hsiâo King. 12*s.* 6*d.*

Vol. IV. The Zend-Avesta. Translated by James Darmesteter. Part I. The Vendîdâd. 10*s.* 6*d.*

Vol. V. The Pahlavi Texts. Translated by E. W. West. Part I. The Bundahi*s*, Bahman Ya*s*t, and Shâyast lâ-shâyast. 12*s.* 6*d.*

Vols. VI and IX. The Qur'ân. Parts I and II. Translated by E. H. Palmer. 21*s.*

Vol. VII. The Institutes of Vish*n*u. Translated by Julius Jolly. 10*s.* 6*d.*

Vol. VIII. The Bhagavadgîtâ, with The Sanatsugâtîya, and The Anugîtâ. Translated by Kâshinâth Trimbak Telang. 10*s.* 6*d.*

Vol. X. The Dhammapada, translated from Pâli by F. Max Müller; and The Sutta-Nipâta, translated from Pâli by V. Fausböll; being Canonical Books of the Buddhists. 10*s.* 6*d.*

Vol. XI. Buddhist Suttas. Translated from Pâli by T. W.
Rhys Davids. 1. The Mahâparinibbâna Suttanta; 2. The Dhamma-*k*akka-ppavattana Sutta; 3. The Tevi*gg*a Suttanta; 4. The Akankheyya Sutta; 5. The *K*etokhila Sutta; 6. The Mahâ-sudassana Suttanta; 7. The Sabbâsava Sutta. 10*s*. 6*d*.

Vol. XII. The *S*atapatha-Brâhma*n*a, according to the Text of the Mâdhyandina School. Translated by Julius Eggeling. Part I. Books I and II. 12*s*. 6*d*.

Vol. XIII. Vinaya Texts. Translated from the Pâli by T. W. Rhys Davids and Hermann Oldenberg. Part I. The Pâtimokkha. The Mahâvagga, I-IV. 10*s*. 6*d*.

Vol. XIV. The Sacred Laws of the Âryas, as taught in the Schools of Âpastamba, Gautama, Vâsish*th*a and Baudhâyana. Translated by Georg Bühler. Part II. Vâsish*th*a and Baudhâyana. 10*s*. 6*d*.

Vol. XV. The Upanishads. Translated by F. Max Müller. Part II. The Ka*th*a-upanishad, The Mu*nd*aka-upanishad, The Taittirîyaka-upanishad, The B*ri*hadâra*n*yaka-upanishad, The *S*veta*s*vatara-upanishad, The Pra*sn*a-upanishad, and The Maitrâya*n*a-Brâhma*n*a-upanishad. 10*s*. 6*d*.

Vol. XVI. The Sacred Books of China. The Texts of Confucianism. Translated by James Legge. Part II. The Yî King. 10*s*. 6*d*.

Vol. XVII. Vinaya Texts. Translated from the Pâli by T. W. Rhys Davids and Hermann Oldenberg. Part II. The Mahâvagga, V-X. The *K*ullavagga, I-III. 10*s*. 6*d*.

Vol. XVIII. Pahlavi Texts. Translated by E. W. West. Part II. The Dâ*d*istân-î Dînîk and The Epistles of Mânû*sk*îhar. 12*s*. 6*d*.

Vol. XIX. The Fo-sho-hing-tsan-king. A Life of Buddha by A*s*vaghosha Bodhisattva, translated from Sanskrit into Chinese by Dharmaraksha, A.D. 420, and from Chinese into English by Samuel Beal. 10*s*. 6*d*.

Vol. XX. Vinaya Texts. Translated from the Pâli by T. W. Rhys Davids and Hermann Oldenberg. Part III. The *K*ullavagga, IV-XII. 10*s*. 6*d*.

Vol. XXI. The Saddharma-pu*nd*arîka; or, the Lotus of the True Law. Translated by H. Kern. 12*s*. 6*d*.

Vol. XXII. *G*aina-Sûtras. Translated from Prâkrit by Hermann Jacobi. Part I. The Â*k*ârânga-Sûtra. The Kalpa-Sûtra. 10*s*. 6*d*.

Vol. XXIII. The Zend-Avesta. Translated by James Darmesteter. Part II. The Sîrôzahs, Yasts, and Nyâyis. 10s. 6d.

Vol. XXIV. Pahlavi Texts. Translated by E. W. West. Part III. Dînâ-î Maînôg-î Khirad, Sikand-gûmânîk, and Sad-Dar. 10s. 6d.

Second Series.

Vol. XXV. Manu. Translated by Georg Bühler. 21s.

Vol. XXVI. The Satapatha-Brâhmana. Translated by Julius Eggeling. Part II. 12s. 6d.

Vols. XXVII and XXVIII. The Sacred Books of China. The Texts of Confucianism. Translated by James Legge. Parts III and IV. The Lî Kî, or Collection of Treatises on the Rules of Propriety, or Ceremonial Usages. 25s.

Vols. XXIX and XXX. The Grihya-Sûtras, Rules of Vedic Domestic Ceremonies. Translated by Hermann Oldenberg.
       Part I (Vol. XXIX), 12s. 6d. *Just Published*.
       Part II (Vol. XXX). *In the Press*.

Vol. XXXI. The Zend-Avesta. Part III. The Yasna, Visparad, Âfrînagân, and Gâhs. Translated by L. H. Mills. 12s. 6d.

The following Volumes are in the Press:—

Vol. XXXII. Vedic Hymns. Translated by F. Max Müller. Part I.

Vol. XXXIII. Nârada, and some Minor Law-books. Translated by Julius Jolly. [*Preparing*.]

Vol. XXXIV. The Vedânta-Sûtras, with Sankara's Commentary. Translated by G. Thibaut. [*Preparing*.]

   _{*} *The Second Series will consist of Twenty-Four Volumes.*

Clarendon Press Series.

I. ENGLISH, &c.

An Elementary English Grammar and Exercise Book. By O. W. Tancock, M.A. Second Edition. Extra fcap. 8vo. 1s. 6d.

An English Grammar and Reading Book, for Lower Forms in Classical Schools. By O. W. Tancock, M.A. Fourth Edition. Extra fcap. 8vo. 3s. 6d.

Typical Selections from the best English Writers, with Introductory Notices. Second Edition. In 2 vols. Extra fcap. 8vo. 3s. 6d. each.
    Vol. I. Latimer to Berkeley.   Vol. II. Pope to Macaulay.

Shairp (J. C., LL.D.). Aspects of Poetry; being Lectures delivered at Oxford. Crown 8vo. 10s. 6d.

A Book for the Beginner in Anglo-Saxon. By John Earle, M.A. Third Edition. Extra fcap. 8vo. 2s. 6d.

An Anglo-Saxon Reader. In Prose and Verse. With Grammatical Introduction, Notes, and Glossary. By Henry Sweet, M.A. Fourth Edition, Revised and Enlarged. Extra fcap. 8vo. 8s. 6d.

A Second Anglo-Saxon Reader. By the same Author. Extra fcap. 8vo. 4s. 6d.

An Anglo-Saxon Primer, with Grammar, Notes, and Glossary. By the same Author. Second Edition. Extra fcap. 8vo. 2s. 6d.

Old English Reading Primers; edited by Henry Sweet, M.A.
    I. Selected Homilies of Ælfric. Extra fcap. 8vo., stiff covers, 1s. 6d.
    II. Extracts from Alfred's Orosius. Extra fcap. 8vo., stiff covers, 1s. 6d.

First Middle English Primer, with Grammar and Glossary. By the same Author. Extra fcap. 8vo. 2s.

Second Middle English Primer. Extracts from Chaucer, with Grammar and Glossary. By the same Author. Extra fcap. 8vo. 2s.

A Concise Dictionary of Middle English, from A.D. 1150 to 1580. By A. L. Mayhew, M.A., and W. W. Skeat, Litt.D. Crown 8vo. half roan, 7s. 6d.

A Handbook of Phonetics, including a Popular Exposition of the Principles of Spelling Reform. By H. Sweet, M.A. Ext. fcap. 8vo. 4s. 6d.

Elementarbuch des Gesprochenen Englisch. Grammatik, Texte und Glossar. Von Henry Sweet. *Second Edition.* Extra fcap. 8vo., stiff covers, 2*s.* 6*d.*

History of English Sounds from the earliest Period. With full Word-Lists. By Henry Sweet, M.A. Demy 8vo. 14*s.*

Principles of English Etymology. First Series. *The Native Element.* By W. W. Skeat, Litt.D. Crown 8vo. 9*s.*

The Philology of the English Tongue. By J. Earle, M.A. Fourth Edition. Extra fcap. 8vo. 7*s.* 6*d.*

An Icelandic Primer, with Grammar, Notes, and Glossary. By Henry Sweet, M.A. Extra fcap. 8vo. 3*s.* 6*d.*

An Icelandic Prose Reader, with Notes, Grammar, and Glossary. By G. Vigfússon, M.A., and F. York Powell, M.A. Ext. fcap. 8vo. 10*s.* 6*d.*

The Ormulum; with the Notes and Glossary of Dr. R. M. White. Edited by R. Holt, M.A. 1878. 2 vols. Extra fcap. 8vo. 21*s.*

Specimens of Early English. A New and Revised Edition. With Introduction, Notes, and Glossarial Index.

    Part I. By R. Morris, LL.D. From Old English Homilies to King Horn (A.D. 1150 to A.D 1300). Second Edition. Extra fcap. 8vo. 9*s.*

    Part II. By R. Morris, LL.D., and W. W. Skeat, Litt.D. From Robert of Gloucester to Gower (A.D. 1298 to A.D. 1393). Third Edition. Extra fcap. 8vo. 7*s.* 6*d.*

Specimens of English Literature, from the 'Ploughmans Crede' to the 'Shepheardes Calender' (A.D. 1394 to A.D. 1579). With Introduction, Notes, and Glossarial Index. By W. W. Skeat, Litt.D. Fourth Edition. Extra fcap. 8vo. 7*s.* 6*d.*

The Vision of William concerning Piers the Plowman, in three Parallel Texts; together with *Richard the Redeless.* By William Langland (about 1362–1399 A.D.). Edited from numerous Manuscripts, with Preface, Notes, and a Glossary, by W. W. Skeat, Litt.D. 2 vols. 8vo. 31*s.* 6*d.*

The Vision of William concerning Piers the Plowman, by William Langland. Edited, with Notes, by W. W. Skeat, Litt.D. Fourth Edition Extra fcap. 8vo. 4*s.* 6*d.*

Chaucer. I. *The Prologue to the Canterbury Tales;* the Knightes Tale; The Nonne Prestes Tale. Edited by R. Morris, LL.D. Sixty-sixth thousand. Extra fcap. 8vo. 2*s.* 6*d.*

    —— II. *The Prioresses Tale; Sir Thopas; The Monkes Tale; The Clerkes Tale; The Squieres Tale*, &c. Edited by W. W. Skeat, Litt.D. Third Edition. Extra fcap. 8vo. 4*s.* 6*d.*

Chaucer. III. *The Tale of the Man of Lawe;* The Pardoneres Tale; The Second Nonnes Tale; The Chanouns Yemannes Tale. By the same Editor. *New Edition, Revised.* Extra fcap. 8vo. 4*s.* 6*d.*

—— IV. *Minor Poems.* By the same Editor. Extra fcap. 8vo. *Just ready.*

Gamelyn, The Tale of. Edited with Notes, Glossary, &c., by W. W. Skeat, Litt.D. Extra fcap. 8vo. Stiff covers, 1*s.* 6*d.*

Minot (Laurence). Poems. Edited, with Introduction and Notes, by Joseph Hall, M.A., Head Master of the Hulme Grammar School, Manchester. Extra fcap. 8vo. 4*s.* 6*d.*

Spenser's Faery Queene. Books I and II. Designed chiefly for the use of Schools. With Introduction and Notes by G. W. Kitchin, D.D., and Glossary by A. L. Mayhew, M.A. Extra fcap. 8vo. 2*s.* 6*d.* each.

Hooker. Ecclesiastical Polity, Book I. Edited by R. W. Church, M.A. Second Edition. Extra fcap. 8vo. 2*s.*

OLD ENGLISH DRAMA.

The Pilgrimage to Parnassus with *The Two Parts of the Return from Parnassus.* Three Comedies performed in St. John's College, Cambridge, A.D. MDXCVII–MDCI. Edited from MSS. by the Rev. W. D. Macray, M.A., F.S.A. Medium 8vo. Bevelled Boards, Gilt top, 8*s.* 6*d.*

Marlowe and Greene. Marlowe's Tragical History of Dr. Faustus, and *Greene's Honourable History of Friar Bacon and Friar Bungay.* Edited by A. W. Ward, M.A. *New and Enlarged Edition.* Extra fcap. 8vo. 6*s.* 6*d.*

Marlowe. Edward II. With Introduction, Notes, &c. By O. W. Tancock, M.A. Extra fcap. 8vo. Paper covers, 2*s.* Cloth 3*s.*

SHAKESPEARE.

Shakespeare. Select Plays. Edited by W. G. Clark, M.A., and W. Aldis Wright, M.A. Extra fcap. 8vo. stiff covers.

The Merchant of Venice. 1*s.*	Macbeth. 1*s.* 6*d.*
Richard the Second. 1*s.* 6*d.*	Hamlet. 2*s.*

Edited by W. Aldis Wright, M.A.

The Tempest. 1*s.* 6*d.*	Midsummer Night's Dream. 1*s.* 6*d.*
As You Like It. 1*s.* 6*d.*	Coriolanus. 2*s.* 6*d.*
Julius Cæsar. 2*s.*	Henry the Fifth. 2*s.*
Richard the Third. 2*s.* 6*d.*	Twelfth Night. 1*s.* 6*d.*
King Lear. 1*s.* 6*d.*	King John. 1*s.* 6*d.*

Shakespeare as a Dramatic Artist; a popular Illustration of the Principles of Scientific Criticism. By R. G. Moulton, M.A. Crown 8vo. 5*s.*

Bacon. I. *Advancement of Learning.* Edited by W. Aldis Wright, M.A. Third Edition. Extra fcap. 8vo. 4*s.* 6*d.*

—— II. *The Essays.* With Introduction and Notes. By S. H. Reynolds, M.A., late Fellow of Brasenose College. *In Preparation.*

Milton. I. *Areopagitica.* With Introduction and Notes. By John W. Hales, M.A. Third Edition. Extra fcap. 8vo. 3*s.*

—— II. *Poems.* Edited by R. C. Browne, M.A. 2 vols. Fifth Edition. Extra fcap. 8vo. 6*s.* 6*d.* Sold separately, Vol. I. 4*s.*; Vol. II. 3*s.*

In paper covers :—

Lycidas, 3*d.*   L'Allegro, 3*d.*   Il Penseroso, 4*d.*   Comus, 6*d.*

—— III. *Paradise Lost.* Book I. Edited by H. C. Beeching. Extra fcap. 8vo. stiff cover, 1*s.* 6*d.*; in white Parchment, 3*s.* 6*d.*

—— IV. *Samson Agonistes.* Edited with Introduction and Notes by John Churton Collins. Extra fcap. 8vo. stiff covers, 1*s.*

Bunyan. I. *The Pilgrim's Progress, Grace Abounding, Relation of the Imprisonment of Mr. John Bunyan.* Edited, with Biographical Introduction and Notes, by E. Venables, M.A. 1879. Extra fcap. 8vo. 5*s.* In ornamental Parchment, 6*s.*

—— II. *Holy War, &c.* Edited by E. Venables, M.A. In the Press.

Clarendon. *History of the Rebellion.* Book VI. Edited by T. Arnold, M.A. Extra fcap. 8vo. 4*s.* 6*d.*

Dryden. *Select Poems.* Stanzas on the Death of Oliver Cromwell; Astræa Redux; Annus Mirabilis; Absalom and Achitophel; Religio Laici; The Hind and the Panther. Edited by W. D. Christie, M.A. Second Edition. Extra fcap. 8vo. 3*s.* 6*d.*

Locke's Conduct of the Understanding. Edited, with Introduction, Notes, &c., by T. Fowler, D.D. Second Edition. Extra fcap. 8vo. 2*s.*

Addison. *Selections from Papers in the Spectator.* With Notes. By T. Arnold, M.A. Extra fcap. 8vo. 4*s.* 6*d.* In ornamental Parchment, 6*s.*

Steele. *Selections from the Tatler, Spectator, and Guardian.* Edited by Austin Dobson. Extra fcap. 8vo. 4*s.* 6*d.* In white Parchment, 7*s.* 6*d.*

Pope. With Introduction and Notes. By Mark Pattison, B.D.

—— I. *Essay on Man.* Extra fcap. 8vo. 1*s.* 6*d.*

—— II. *Satires and Epistles.* Extra fcap. 8vo. 2*s.*

Parnell. *The Hermit.* Paper covers, 2*d.*

Gray. *Selected Poems.* Edited by Edmund Gosse. Extra fcap. 8vo. Stiff covers, 1*s.* 6*d.* In white Parchment, 3*s.*

—— *Elegy and Ode on Eton College.* Paper covers, 2*d.*

Goldsmith. Selected Poems. Edited, with Introduction and Notes, by Austin Dobson. Extra fcap. 8vo. 3*s.* 6*d.* In white Parchment, 4*s.* 6*d.*

—— *The Traveller.* With Notes by G. Birkbeck Hill, D.C.L. Extra fcap. 8vo. *Just ready.*

—— *The Deserted Village.* Paper covers, 2*d.*

Johnson. I. *Rasselas.* Edited, with Introduction and Notes, by G. Birkbeck Hill, D.C.L. Extra fcap. 8vo. Bevelled boards, 3*s.* 6*d.* In white Parchment, 4*s.* 6*d.*

—— II. *Rasselas; Lives of Dryden and Pope.* Edited by Alfred Milnes, M.A. (London). Extra fcap. 8vo. 4*s.* 6*d.*
    Lives of Dryden and Pope. Stiff covers, 2*s.* 6*d.*

—— III. *Life of Milton.* Edited, with Notes, etc., by C. H. Firth, M.A. Extra fcap. 8vo. *cloth*, 2*s.* 6*d.* *Stiff cover*, 1*s.* 6*d.*

—— IV. *Vanity of Human Wishes.* With Notes, by E. J. Payne, M.A. Paper covers, 4*d.*

—— V. *Wit and Wisdom of Samuel Johnson.* Edited by G. Birkbeck Hill, D.C.L. Crown 8vo. 7*s.* 6*d.*

—— VI. *Boswell's Life of Johnson. With the Journal of a Tour to the Hebrides.* Edited, with copious Notes, Appendices, and Index, by G. Birkbeck Hill, D.C.L., Pembroke College. With Portraits and Facsimiles. 6 vols. Medium 8vo. *Half bound*, 3*l.* 3*s.*

Cowper. Edited, with Life, Introductions, and Notes, by H. T. Griffith, B.A.

—— I. *The Didactic Poems of* 1782, with Selections from the Minor Pieces, A.D. 1779-1783. Extra fcap. 8vo. 3*s.*

—— II. *The Task, with Tirocinium*, and Selections from the Minor Poems, A.D. 1784-1799. Second Edition. Extra fcap. 8vo. 3*s.*

Burke. Select Works. Edited, with Introduction and Notes, by E. J. Payne, M.A.

—— I. *Thoughts on the Present Discontents; the two Speeches on America.* Second Edition. Extra fcap. 8vo. 4*s.* 6*d.*

—— II. *Reflections on the French Revolution.* Second Edition. Extra fcap. 8vo. 5*s.*

—— III. *Four Letters on the Proposals for Peace with the Regicide Directory of France.* Second Edition. Extra fcap. 8vo. 5*s.*

Keats. Hyperion, Book I. With Notes by W. T. Arnold, B.A. Paper covers, 4*d.*

Byron. Childe Harold. Edited, with Introduction and Notes, by H. F. Tozer, M.A. Extra fcap. 8vo. 3*s.* 6*d.* In white Parchment, 5*s.*

Scott. Lay of the Last Minstrel. Edited with Preface and Notes by W. Minto, M.A. With Map. Extra fcap. 8vo. Stiff covers, 2s. Ornamental Parchment, 3s. 6d.

—— *Lay of the Last Minstrel.* Introduction and Canto I, with Preface and Notes, by the same Editor. 6d.

II. LATIN.

Rudimenta Latina. Comprising Accidence, and Exercises of a very Elementary Character, for the use of Beginners. By John Barrow Allen, M.A. Extra fcap. 8vo. 2s.

An Elementary Latin Grammar. By the same Author. Fifty-Seventh Thousand. Extra fcap. 8vo. 2s. 6d.

A First Latin Exercise Book. By the same Author. Fourth Edition. Extra fcap. 8vo. 2s. 6d.

A Second Latin Exercise Book. By the same Author. Extra fcap. 8vo. 3s. 6d.

A Key to First and Second Latin Exercise Books, in one volume, price 5s. Supplied to *Teachers only* on application to the Secretary of the Clarendon Press.

Reddenda Minora, or Easy Passages, Latin and Greek, for Unseen Translation. For the use of Lower Forms. Composed and selected by C. S. Jerram, M.A. Extra fcap. 8vo. 1s. 6d.

Anglice Reddenda, or Extracts, Latin and Greek, for Unseen Translation. By C. S. Jerram, M.A. Third Edition, Revised and Enlarged. Extra fcap. 8vo. 2s. 6d.

Anglice Reddenda. Second Series. By the same Author. Extra fcap. 8vo. 3s.

Passages for Translation into Latin. For the use of Passmen and others. Selected by J. Y. Sargent, M.A. Seventh Edition. Extra fcap. 8vo. 2s. 6d.

Exercises in Latin Prose Composition; with Introduction, Notes, and Passages of Graduated Difficulty for Translation into Latin. By G. G. Ramsay, M.A., LL.D. Second Edition. Extra fcap. 8vo. 4s. 6d.

Hints and Helps for Latin Elegiacs. By H. Lee-Warner, M.A. Extra fcap. 8vo. 3s. 6d.

First Latin Reader. By T. J. Nunns, M.A. Third Edition. Extra fcap. 8vo. 2s.

Caesar. The Commentaries (for Schools). With Notes and Maps. By Charles E. Moberly, M.A.

The Gallic War. Second Edition. Extra fcap. 8vo. 4s. 6d.
The Gallic War. Books I, II. Extra fcap. 8vo. *Just ready.*
The Civil War. Extra fcap. 8vo. 3s. 6d.
The Civil War. Book I. Second Edition. Extra fcap. 8vo. 2s.

Cicero. Speeches against Catilina. By E. A. Upcott, M.A., Assistant Master in Wellington College. In one or two Parts. Extra fcap. 8vo. 2s. 6d.

—— *Selection of interesting and descriptive passages.* With Notes. By Henry Walford, M.A. In three Parts. Extra fcap. 8vo. 4s. 6d.
Each Part separately, limp, 1s. 6d.
Part I. Anecdotes from Grecian and Roman History. Third Edition.
Part II. Omens and Dreams: Beauties of Nature. Third Edition.
Part III. Rome's Rule of her Provinces. Third Edition.

—— *De Senectute.* Edited, with Introduction and Notes, by L. Huxley, M.A. In one or two Parts. Extra fcap. 8vo. 2s.

—— *Selected Letters* (for Schools). With Notes. By the late C. E. Prichard, M.A., and E. R. Bernard, M.A. Second Edition. Extra fcap. 8vo. 3s.

—— *Select Orations* (for Schools). In Verrem I. De Imperio Gn. Pompeii. Pro Archia. Philippica IX. With Introduction and Notes by J. R. King, M.A. Second Edition. Extra fcap. 8vo. 2s. 6d.

—— *In Q. Caecilium Divinatio,* and *In C. Verrem Actio Prima.* With Introduction and Notes, by J. R. King, M.A. Extra fcap. 8vo. limp, 1s. 6d.

—— *Speeches against Catilina.* With Introduction and Notes, by E. A. Upcott, M.A. In one or two Parts. Extra fcap. 8vo. 2s. 6d.

Cornelius Nepos. With English Notes. By Oscar Browning, M.A. Third Edition. Revised by W. R. Inge, M.A. (In one or two Parts.) Extra fcap. 8vo. 3s.

Horace. Selected Odes. With Notes for the use of a Fifth Form. By E. C. Wickham, M.A. In one or two Parts. Extra fcap. 8vo. cloth, 2s.

Livy. Selections (for Schools). With Notes and Maps. By H. Lee-Warner, M.A. Extra fcap. 8vo. In Parts, limp, each 1s. 6d.
Part I. The Caudine Disaster. Part II. Hannibal's Campaign in Italy. Part III. The Macedonian War.

—— Books V–VII. With Introduction and Notes. By A. R. Cluer, B.A. Second Edition. Revised by P. E. Matheson, M.A. (In one or two Parts.) Extra fcap. 8vo. 5s.

—— Books XXI, XXII, and XXIII. With Introduction and Notes. By M. T. Tatham, M.A. Extra fcap. 8vo. 4s. 6d.

—— Book XXII. By the same Editor. Extra fcap. 8vo. *Just ready.*

Ovid. Selections for the use of Schools. With Introductions and Notes, and an Appendix on the Roman Calendar. By W. Ramsay, M.A. Edited by G. G. Ramsay, M.A. Third Edition. Extra fcap. 8vo. 5s. 6d.

Ovid. Tristia. Book I. The Text revised, with an Introduction and Notes. By S. G. Owen, B.A. Extra fcap. 8vo. 3*s.* 6*d.*

Plautus. Captivi. Edited by W. M. Lindsay, M.A. Extra fcap. 8vo. (In one or two Parts.) 2*s.* 6*d.*

—— *The Trinummus.* With Notes and Introductions. (Intended for the Higher Forms of Public Schools.) By C. E. Freeman, M.A., and A. Sloman, M.A. Extra fcap. 8vo. 3*s.*

Pliny. Selected Letters (for Schools). With Notes. By the late C. E. Prichard, M.A., and E. R. Bernard, M.A. Extra fcap. 8vo. 3*s.*

Sallust. With Introduction and Notes. By W. W. Capes, M.A. Extra fcap. 8vo. 4*s.* 6*d.*

Tacitus. The Annals. Books I–IV. Edited, with Introduction and Notes (for the use of Schools and Junior Students), by H. Furneaux, M.A. Extra fcap. 8vo. 5*s.*

—— *The Annals.* Book I. With Introduction and Notes, by the same Editor. Extra fcap. 8vo. limp, 2*s.*

Terence. Andria. With Notes and Introductions. By C. E. Freeman, M.A., and A. Sloman, M.A. Extra fcap. 8vo. 3*s.*

—— *Adelphi.* With Notes and Introductions. (Intended for the Higher Forms of Public Schools.) By A. Sloman, M.A. Extra fcap. 8vo. 3*s.*

—— *Phormio.* With Notes and Introductions. By A. Sloman, M.A. Extra fcap. 8vo. 3*s.*

Tibullus and Propertius. Selections. Edited by G. G. Ramsay, M.A. Extra fcap. 8vo. (In one or two vols.) 6*s.*

Virgil. With Introduction and Notes. By T. L. Papillon, M.A. Two vols. Crown 8vo. 10*s.* 6*d.* The Text separately, 4*s.* 6*d.*

—— *Bucolics.* Edited by C. S. Jerram, M.A. In one or two Parts. Extra fcap. 8vo. 2*s.* 6*d.*

—— *Aeneid* I. With Introduction and Notes, by C. S. Jerram, M.A. Extra fcap. 8vo. limp, 1*s.* 6*d.*

—— *Aeneid* IX. Edited, with Introduction and Notes, by A. E. Haigh, M.A., late Fellow of Hertford College, Oxford. Extra fcap. 8vo. limp, 1*s.* 6*d.* In two Parts, 2*s.*

Avianus, The Fables of. Edited, with Prolegomena, Critical Apparatus, Commentary, etc. By Robinson Ellis, M.A., LL.D. Demy 8vo. 8*s.* 6*d.*

Catulli Veronensis Liber. Iterum recognovit, apparatum criticum prolegomena appendices addidit, Robinson Ellis, A.M. 1878. Demy 8vo. 16*s.*

—— *A Commentary on Catullus.* By Robinson Ellis, M.A. 1876. Demy 8vo. 16*s.*

Catulli Veronensis Carmina Selecta, secundum recognitionem
Robinson Ellis, A.M. Extra fcap. 8vo. 3*s*. 6*d*.

Cicero de Oratore. With Introduction and Notes. By A. S. Wilkins, Litt. D.
Book I. Second Edition. 1888. 8vo. 7*s*. 6*d*. Book II. 1881. 8vo. 5*s*.

—— *Philippic Orations.* With Notes. By J. R. King, M.A.
Second Edition. 1879. 8vo. 10*s*. 6*d*.

—— *Select Letters.* With English Introductions, Notes, and Appendices. By Albert Watson, M.A. Third Edition. Demy 8vo. 18*s*.

—— *Select Letters.* Text. By the same Editor. Second Edition. Extra fcap. 8vo. 4*s*.

—— *pro Cluentio.* With Introduction and Notes. By W. Ramsay, M.A. Edited by G. G. Ramsay, M.A. 2nd Ed. Ext. fcap. 8vo. 3*s*. 6*d*.

Horace. With a Commentary. Volume I. The Odes, Carmen Seculare, and Epodes. By Edward C. Wickham, M.A. Second Edition. 1877. Demy 8vo. 12*s*.

—— A reprint of the above, in a size suitable for the use of Schools. In one or two Parts. Extra fcap. 8vo. 6*s*.

Livy, Book I. With Introduction, Historical Examination, and Notes. By J. R. Seeley, M.A. Second Edition. 1881. 8vo. 6*s*.

Ovid. P. Ovidii Nasonis Ibis. Ex Novis Codicibus edidit, Scholia Vetera Commentarium cum Prolegomenis Appendice Indice addidit, R. Ellis, A.M. 8vo. 10*s*. 6*d*.

Persius. The Satires. With a Translation and Commentary. By John Conington, M.A. Edited by Henry Nettleship, M.A. Second Edition. 1874. 8vo. 7*s*. 6*d*.

Juvenal. XIII Satires. Edited, with Introduction and Notes, by C. H. Pearson, M.A., and Herbert A. Strong, M.A., LL.D., Professor of Latin in Liverpool University College, Victoria University. In two Parts. Crown 8vo. Complete, 6*s*.

Also separately, Part I. Introduction, Text, etc., 3*s*. Part II. Notes, 3*s*. 6*d*.

Tacitus. The Annals. Books I-VI. Edited, with Introduction and Notes, by H. Furneaux, M.A. 8vo. 18*s*.

King (J. E., M.A.) and C. Cookson, M.A. The Principles of Sound and Inflexion, as illustrated in the Greek and Latin Languages. 1888. 8vo. 18*s*.

Nettleship (H., M.A.). Lectures and Essays on Subjects connected with Latin Scholarship and Literature. Crown 8vo. 7*s*. 6*d*.

—— *The Roman Satura.* 8vo. sewed, 1*s*.

—— *Ancient Lives of Vergil.* 8vo. sewed, 2*s*.

Papillon (T. L., M.A.). A Manual of Comparative Philology.
Third Edition, Revised and Corrected. 1882. Crown 8vo. 6s.

Pinder (North, M.A.). Selections from the less known Latin Poets. 1869. 8vo. 15s.

Sellar (W. Y., M.A.). Roman Poets of the Augustan Age.
VIRGIL. New Edition. 1883. Crown 8vo. 9s.

—— *Roman Poets of the Republic.* New Edition, Revised and Enlarged. 1881. 8vo. 14s.

Wordsworth (J., M.A.). Fragments and Specimens of Early Latin. With Introductions and Notes. 1874. 8vo. 18s.

III. GREEK.

A Greek Primer, for the use of beginners in that Language.
By Charles Wordsworth, D.C.L. Seventh Edition. Extra fcap. 8vo. 1s. 6d.

A Greek Testament Primer. An Easy Grammar and Reading Book for the use of Students beginning Greek. By the Rev. E. Miller, M.A. Extra fcap. 8vo. 3s. 6d.

Easy Greek Reader. By Evelyn Abbott, M.A. In one or two Parts. Extra fcap. 8vo. 3s.

Graecae Grammaticae Rudimenta in usum Scholarum. Auctore Carolo Wordsworth, D.C.L. Nineteenth Edition, 1882. 12mo. 4s.

A Greek-English Lexicon, abridged from Liddell and Scott's 4to. edition, chiefly for the use of Schools. Twenty-first Edition. 1886. Square 12mo. 7s. 6d.

Greek Verbs, Irregular and Defective. By W. Veitch. Fourth Edition. Crown 8vo. 10s. 6d.

The Elements of Greek Accentuation (for Schools): abridged from his larger work by H. W. Chandler, M.A. Extra fcap. 8vo. 2s. 6d.

A SERIES OF GRADUATED GREEK READERS:—

First Greek Reader. By W. G. Rushbrooke, M.L. Second Edition. Extra fcap. 8vo. 2s. 6d.

Second Greek Reader. By A. M. Bell, M.A. Extra fcap. 8vo. 3s. 6d.

Fourth Greek Reader; being Specimens of Greek Dialects. With Introductions, etc. By W. W. Merry, D.D. Extra fcap. 8vo. 4s. 6d.

Fifth Greek Reader. Selections from Greek Epic and Dramatic Poetry, with Introductions and Notes. By Evelyn Abbott, M.A. Extra fcap. 8vo. 4s. 6d.

The Golden Treasury of Ancient Greek Poetry: being a Collection of the finest passages in the Greek Classic Poets, with Introductory Notices and Notes. By R. S. Wright, M.A. Extra fcap. 8vo. 8s. 6d.

A Golden Treasury of Greek Prose, being a Collection of the finest passages in the principal Greek Prose Writers, with Introductory Notices and Notes. By R. S. Wright, M.A., and J. E. L. Shadwell, M.A. Extra fcap. 8vo. 4s. 6d.

Aeschylus. Prometheus Bound (for Schools). With Introduction and Notes, by A. O. Prickard, M.A. Second Edition. Extra fcap. 8vo. 2s.

—— *Agamemnon.* With Introduction and Notes, by Arthur Sidgwick, M.A. Third Edition. In one or two parts. Extra fcap. 8vo. 3s.

—— *Choephoroi.* With Introduction and Notes by the same Editor. Extra fcap. 8vo. 3s.

—— *Eumenides.* With Introduction and Notes, by the same Editor. In one or two Parts. Extra fcap. 8vo. 3s.

Aristophanes. In Single Plays. Edited, with English Notes, Introductions, &c., by W. W. Merry, D.D. Extra fcap. 8vo.
I. The Clouds, Second Edition, 2s.
II. The Acharnians, Third Edition. In one or two parts, 3s.
III. The Frogs, Second Edition. In one or two parts, 3s.
IV. The Knights. In one or two parts, 3s.

Cebes. Tabula. With Introduction and Notes. By C. S. Jerram, M.A. Extra fcap. 8vo. 2s. 6d.

Demosthenes. Orations against Philip. With Introduction and Notes, by Evelyn Abbott, M.A., and P. E. Matheson, M.A. Vol. I. Philippic I. Olynthiacs I–III. In one or two Parts. Extra fcap. 8vo. 3s.

Euripides. Alcestis (for Schools). By C. S. Jerram, M.A. Extra fcap. 8vo. 2s. 6d.

—— *Hecuba.* With Notes by C. H. Russell. *In the Press.*

—— *Helena.* Edited, with Introduction, Notes, etc., for Upper and Middle Forms. By C. S. Jerram, M.A. Extra fcap. 8vo. 3s.

—— *Heracleidae.* Edited with Introduction and Notes by C. S. Jerram, M.A. Extra fcap. 8vo. 3s.

—— *Iphigenia in Tauris.* Edited, with Introduction, Notes, etc., for Upper and Middle Forms. By C. S. Jerram, M.A. Extra fcap. 8vo. cloth, 3s.

—— *Medea.* By C. B. Heberden, M.A. In one or two Parts. Extra fcap. 8vo. 2s.

Herodotus, Book IX. Edited, with Notes, by Evelyn Abbott, M.A. In one or two Parts. Extra fcap. 8vo. 3s.

Herodotus, Selections from. Edited, with Introduction, Notes, and a Map, by W. W. Merry, D.D. Extra fcap. 8vo. 2s. 6d.

Homer. Odyssey, Books I-XII (for Schools). By W. W. Merry, D.D. Fortieth Thousand. (In one or two Parts.) Extra fcap. 8vo. 5s.
Books I, and II, *separately*. each 1s. 6d.

—— *Odyssey*, Books XIII-XXIV (for Schools). By the same Editor. Second Edition. Extra fcap. 8vo. 5s.

—— *Iliad*, Book I (for Schools). By D. B. Monro, M.A. Second Edition. Extra fcap. 8vo. 2s.

—— *Iliad*, Books I-XII (for Schools). With an Introduction, a brief Homeric Grammar, and Notes. By D. B. Monro, M.A. Second Edition. Extra fcap. 8vo. 6s.

—— *Iliad*, Books VI and XXI. With Introduction and Notes. By Herbert Hailstone, M.A. Extra fcap. 8vo. 1s. 6d. each.

Lucian. Vera Historia (for Schools). By C. S. Jerram, M.A. Second Edition. Extra fcap. 8vo. 1s. 6d.

Lysias. Epitaphios. Edited, with Introduction and Notes, by F. J. Snell, B.A. (In one or two Parts.) Extra fcap. 8vo. 2s.

Plato. Meno. With Introduction and Notes. By St. George Stock, M.A., Pembroke College. (In one or two Parts.) Extra fcap. 8vo. 2s. 6d.

Plato. The Apology. With Introduction and Notes. By St. George Stock, M.A. (In one or two Parts.) Extra fcap. 8vo. 2s. 6d.

Sophocles. For the use of Schools. Edited with Introductions and English Notes By Lewis Campbell, M.A., and Evelyn Abbott, M.A. *New and Revised Edition.* 2 Vols. Extra fcap. 8vo. 10s. 6d.
Sold separately, Vol. I, Text, 4s. 6d.; Vol. II, Explanatory Notes, 6s.

Sophocles. In Single Plays, with English Notes, &c. By Lewis Campbell, M.A., and Evelyn Abbott, M.A. Extra fcap. 8vo. limp.
Oedipus Tyrannus, Philoctetes. New and Revised Edition, 2s. each.
Oedipus Coloneus, Antigone, 1s. 9d. each.
Ajax, Electra, Trachiniae, 2s. each.

—— *Oedipus Rex:* Dindorf's Text, with Notes by the present Bishop of St. David's. Extra fcap. 8vo. limp, 1s. 6d.

Theocritus (for Schools). With Notes. By H. Kynaston, D.D. (late Snow). Third Edition. Extra fcap. 8vo. 4s. 6d.

Xenophon. Easy Selections (for Junior Classes). With a Vocabulary, Notes, and Map. By J. S. Phillpotts, B.C.L., and C. S. Jerram, M.A. Third Edition. Extra fcap. 8vo. 3s. 6d.

—— *Selections* (for Schools). With Notes and Maps. By J. S. Phillpotts, B.C.L. Fourth Edition. Extra fcap. 8vo. 3s. 6d.

—— *Anabasis*, Book I. Edited for the use of Junior Classes and Private Students. With Introduction, Notes, etc. By J. Marshall, M.A., Rector of the Royal High School, Edinburgh. Extra fcap. 8vo. 2s. 6d.

Xenophon. Anabasis, Book II. With Notes and Map. By
C. S. Jerram, M.A. Extra fcap. 8vo. 2s.

—— *Anabasis*, Book III. Edited with Introduction, Analysis,
Notes, etc., by J. Marshall, M.A. Extra fcap. 8vo. 2s. 6d.

—— *Cyropaedia*, Book I. With Introduction and Notes by
C. Bigg, D.D. Extra fcap. 8vo. 2s.

—— *Cyropaedia*, Books IV and V. With Introduction and
Notes by C. Bigg, D.D. Extra fcap. 8vo. 2s. 6d.

—— *Hellenica*, Books I, II. With Introductions and Notes
by G. E. Underhill, M.A., Fellow and Tutor of Magdalen College. Extra
fcap. 8vo. *cloth*, 3s.

Aristotle's Politics. With an Introduction, Essays, and Notes.
By W. L. Newman, M.A., Fellow of Balliol College. Vols. I and II.
Medium 8vo. 28s.

Aristotle. On the History of the Process by which the Aristotelian Writings arrived at their present form. An Essay by Richard Shute,
M.A., late Student of Christ Church; with a Brief Memoir of the Author.
8vo. 7s. 6d.

Aristotelian Studies. I. On the Structure of the Seventh
Book of the Nicomachean Ethics. By J. C. Wilson, M.A. 8vo. stiff, 5s.

Aristotelis Ethica Nicomachea, ex recensione Immanuelis
Bekkeri. Crown 8vo. 5s.

Demosthenes and Aeschines. The Orations of Demosthenes
and Æschines on the Crown. With Introductory Essays and Notes. By
G. A. Simcox, M.A., and W. H. Simcox, M.A. 1872. 8vo. 12s.

*Head (Barclay V.). Historia Numorum: A Manual of Greek
Numismatics.* Royal 8vo. half-bound. 2l. 2s.

Hicks (E. L., M.A.). A Manual of Greek Historical Inscriptions. Demy 8vo. 10s. 6d.

Homer. Odyssey, Books I–XII. Edited with English Notes,
Appendices, etc. By W. W. Merry, D.D., and the late James Riddell, M.A.
1886. Second Edition. Demy 8vo. 16s.

Homer. A Grammar of the Homeric Dialect. By D. B. Monro,
M.A. Demy 8vo. 10s. 6d.

Polybius. Selections from Polybius. Edited by J. L. Strachan-
Davidson, M.A., Fellow and Tutor of Balliol College. With three Maps.
Medium 8vo. buckram, 21s.

Sophocles. The Plays and Fragments. With English Notes
and Introductions, by Lewis Campbell, M.A. 2 vols.
    Vol. I. Oedipus Tyrannus. Oedipus Coloneus. Antigone. 8vo. 16s.
    Vol. II. Ajax. Electra. Trachiniae. Philoctetes. Fragments. 8vo. 16s.

IV. FRENCH AND ITALIAN.

Brachet's Etymological Dictionary of the French Language.
Translated by G. W. Kitchin, D.D. Third Edition. Crown 8vo. 7s. 6d.

—— *Historical Grammar of the French Language.* Translated by G. W. Kitchin, D.D. Fourth Edition. Extra fcap. 8vo. 3s. 6d.

Works by GEORGE SAINTSBURY, M.A.

Primer of French Literature. Extra fcap. 8vo. 2s.

Short History of French Literature. Crown 8vo. 10s. 6d.

Specimens of French Literature, from Villon to Hugo. Crown 8vo. 9s.

MASTERPIECES OF THE FRENCH DRAMA.

Corneille's Horace. Edited, with Introduction and Notes, by George Saintsbury, M.A. Extra fcap. 8vo. 2s. 6d.

Molière's Les Précieuses Ridicules. Edited, with Introduction and Notes, by Andrew Lang, M.A. Extra fcap. 8vo. 1s. 6d.

Racine's Esther. Edited, with Introduction and Notes, by George Saintsbury, M.A. Extra fcap. 8vo. 2s.

Beaumarchais' Le Barbier de Séville. Edited, with Introduction and Notes, by Austin Dobson. Extra fcap. 8vo. 2s. 6d.

Voltaire's Mérope. Edited, with Introduction and Notes, by George Saintsbury. Extra fcap. 8vo. cloth, 2s.

Musset's On ne badine pas avec l'Amour, and *Fantasio.* Edited, with Prolegomena, Notes, etc., by Walter Herries Pollock. Extra fcap. 8vo. 2s.

The above six Plays may be had in ornamental case, and bound in Imitation Parchment, price 12s. 6d.

Perrault's Popular Tales. Edited from the Original Editions, with Introduction, etc., by Andrew Lang, M.A. Extra fcap. 8vo., paper boards, 5s. 6d.

Sainte-Beuve. Selections from the Causeries du Lundi. Edited by George Saintsbury, M.A. Extra fcap. 8vo. 2s.

Quinet's Lettres à sa Mère. Selected and edited by George Saintsbury, M.A. Extra fcap. 8vo. 2s.

Gautier, Théophile. Scenes of Travel. Selected and Edited by George Saintsbury, M.A. Extra fcap. 8vo. 2s.

L'Éloquence de la Chaire et de la Tribune Françaises. Edited by Paul Blouët, B.A. Vol. I. Sacred Oratory. Extra fcap. 8vo. 2s. 6d.

Edited by GUSTAVE MASSON, B.A.

Corneille's Cinna. With Notes, Glossary, etc. Extra fcap. 8vo. cloth, 2s. Stiff covers, 1s. 6d.

Louis XIV and his Contemporaries; as described in Extracts from the best Memoirs of the Seventeenth Century. With English Notes, Genealogical Tables, &c. Extra fcap. 8vo. 2s. 6d.

Maistre, Xavier de. Voyage autour de ma Chambre. Ourika, by *Madame de Duras;* Le Vieux Tailleur, by *MM. Erckmann-Chatrian;* La Veillée de Vincennes, by *Alfred de Vigny;* Les Jumeaux de l'Hôtel Corneille, by *Edmond About;* Mésaventures d'un Écolier, by *Rodolphe Töpffer.* Third Edition, Revised and Corrected. Extra fcap. 8vo. 2s. 6d.

—— *Voyage autour de ma Chambre.* Limp, 1s. 6d.

Molière's Les Fourberies de Scapin, and *Racine's Athalie.* With Voltaire's Life of Molière. Extra fcap. 8vo. 2s. 6d.

Molière's Les Fourberies de Scapin. With Voltaire's Life of Molière. Extra fcap. 8vo. stiff covers, 1s. 6d.

Molière's Les Femmes Savantes. With Notes, Glossary, etc. Extra fcap. 8vo. cloth, 2s. Stiff covers, 1s. 6d.

Racine's Andromaque, and *Corneille's Le Menteur.* With Louis Racine's Life of his Father. Extra fcap. 8vo. 2s. 6d.

Regnard's Le Joueur, and *Brueys and Palaprat's Le Grondeur.* Extra fcap. 8vo. 2s. 6d.

Sévigné, Madame de, and her chief Contemporaries, Selections from the Correspondence of. Intended more especially for Girls' Schools. Extra fcap. 8vo. 3s.

Dante. Selections from the Inferno. With Introduction and Notes. By H. B. Cotterill, B.A. Extra fcap. 8vo. 4s. 6d.

Tasso. La Gerusalemme Liberata. Cantos i, ii. With Introduction and Notes. By the same Editor. Extra fcap. 8vo. 2s. 6d.

V. GERMAN.

Scherer (W.). A History of German Literature. Translated from the Third German Edition by Mrs. F. Conybeare. Edited by F. Max Müller. 2 vols. 8vo. 21s.

Max Müller. The German Classics, from the Fourth to the Nineteenth Century. With Biographical Notices, Translations into Modern German, and Notes. By F. Max Müller, M.A. A New Edition, Revised, Enlarged, and Adapted to Wilhelm Scherer's 'History of German Literature,' by F. Lichtenstein. 2 vols. crown 8vo. 21s.

GERMAN COURSE. By HERMANN LANGE.

The Germans at Home; a Practical Introduction to German Conversation, with an Appendix containing the Essentials of German Grammar. Third Edition. 8vo. 2s. 6d.

The German Manual; a German Grammar, Reading Book, and a Handbook of German Conversation. 8vo. 7s. 6d.

Grammar of the German Language. 8vo. 3s. 6d.

German Composition; A Theoretical and Practical Guide to the Art of Translating English Prose into German. Ed. 2. 8vo. 4s. 6d.

German Spelling; A Synopsis of the Changes which it has undergone through the Government Regulations of 1880. Paper covers, 6d.

Lessing's Laokoon. With Introduction, English Notes, etc. By A. Hamann, Phil. Doc., M.A. Extra fcap. 8vo. 4s. 6d.

Schiller's Wilhelm Tell. Translated into English Verse by E. Massie, M.A. Extra fcap. 8vo. 5s.

GERMAN CLASSICS.

With Biographical, Historical, and Critical Introductions, Arguments (to the Dramas), and Complete Commentaries.

Edited by C. A. BUCHHEIM, Phil. Doc. Professor in King's College, London.

Lessing:
  (a) *Nathan der Weise.* A Dramatic Poem. 4s. 6d.
  (b) *Minna von Barnhelm.* A Comedy. 3s. 6d.

Goethe:
  (a) *Egmont.* A Tragedy. 3s.
  (b) *Iphigenie auf Tauris.* A Drama. 3s.

Schiller:
  (a) *Wilhelm Tell.* A Drama. Large Edition. With a Map. 3s. 6d.
  (b) *Wilhelm Tell.* School Edition. With a Map. 2s.
  (c) *Historische Skizzen.* With a Map. 2s. 6d.

Heine:
  (a) *Prosa:* being Selections from his Prose Writings. 4s. 6d.
  (b) *Harzreise.* Cloth, 2s. 6d.; paper covers, 1s. 6d.

Modern German Reader. A Graduated Collection of Extracts from Modern German Authors:—
  Part I. Prose Extracts. With English Notes, a Grammatical Appendix, and a Complete Vocabulary. Fourth Edition. 2s. 6d.
  Part II. Extracts in Prose and Poetry. With English Notes and an Index. Second Edition. 2s. 6d.

Becker (the Historian):
   Friedrich der Grosse. Edited, with Notes, an Historical Introduction, and a Map. 3s. 6d.

Niebuhr:
   Griechische Heroen-Geschichten (Tales of Greek Heroes). Edited, with English Notes and a Vocabulary, by Emma S. Buchheim. Second, Revised Edition. *cloth*, 2s.

An Old High German Primer. With Grammar, Notes, and Glossary. By Joseph Wright, Ph.D. Extra fcap. 8vo. 3s. 6d.

A Middle High German Primer. With Grammar, Notes, and Glossary. By Joseph Wright, Ph.D. Extra fcap. 8vo. 3s. 6d.

VI. MATHEMATICS, PHYSICAL SCIENCE, &c.
By LEWIS HENSLEY, M.A.

Figures made Easy: a first Arithmetic Book. Crown 8vo. 6d.

Answers to the Examples in Figures made Easy, together with two thousand additional Examples, with Answers. Crown 8vo. 1s.

The Scholar's Arithmetic. Crown 8vo. 2s. 6d.

Answers to the Examples in the Scholar's Arithmetic. 1s. 6d.

The Scholar's Algebra. Crown 8vo. 2s. 6d.

Aldis (*W. S., M.A.*). *A Text-Book of Algebra: with Answers to the Examples.* Crown 8vo. 7s. 6d.

Baynes (*R. E., M.A.*). *Lessons on Thermodynamics.* 1878. Crown 8vo. 7s. 6d.

Chambers (*G. F., F.R.A.S.*). *A Handbook of Descriptive Astronomy.* Third Edition. 1877. Demy 8vo. 28s.

Clarke (*Col. A. R., C.B., R.E.*). *Geodesy.* 1880. 8vo. 12s. 6d.

Cremona (*Luigi*). *Elements of Projective Geometry.* Translated by C. Leudesdorf, M.A. 8vo. 12s. 6d.

Donkin. Acoustics. Second Edition. Crown 8vo. 7s. 6d.

Etheridge (*R.*). *Fossils of the British Islands, Stratigraphically arranged.* Part I. PALAEOZOIC. 4to. 1l. 10s. *Just ready.*

Euclid Revised. Containing the Essentials of the Elements of Plane Geometry as given by Euclid in his first Six Books. Edited by R. C. J. Nixon, M.A. Crown 8vo.

      Sold separately as follows,

    Book I. 1s.   Books I, II. 1s. 6d.
    Books I–IV. 3s. 6d.   Books V, VI. 3s.

Euclid.—Geometry in Space. Containing parts of Euclid's Eleventh and Twelfth Books. By the same Editor. Crown 8vo. 3s. 6d.

Galton (Douglas, C.B., F.R.S.). The Construction of Healthy Dwellings. Demy 8vo. 10s. 6d.

Hamilton (Sir R. G. C.), and J. Ball. Book-keeping. New and enlarged Edition. Extra fcap. 8vo. limp cloth, 2s.

Ruled Exercise books adapted to the above may be had, price 2s.

Harcourt (A. G. Vernon, M.A.), and H. G. Madan, M.A. Exercises in Practical Chemistry. Vol. I. Elementary Exercises. Fourth Edition. Crown 8vo. 10s. 6d.

Maclaren (Archibald). A System of Physical Education: Theoretical and Practical. Extra fcap. 8vo. 7s. 6d.

Madan (H. G., M.A.). Tables of Qualitative Analysis. Large 4to. paper, 4s. 6d.

Maxwell (J. Clerk, M.A., F.R.S.). A Treatise on Electricity and Magnetism. Second Edition. 2 vols. Demy 8vo. 1l. 11s. 6d.

—— *An Elementary Treatise on Electricity.* Edited by William Garnett, M.A. Demy 8vo. 7s. 6d.

Minchin (G. M., M.A.). A Treatise on Statics with Applications to Physics. Third Edition, Corrected and Enlarged. Vol. I. *Equilibrium of Coplanar Forces.* 8vo. 9s. Vol. II. *Statics.* 8vo. 16s.

—— *Uniplanar Kinematics of Solids and Fluids.* Crown 8vo. 7s. 6d.

Phillips (John, M.A., F.R.S.). Geology of Oxford and the Valley of the Thames. 1871. 8vo. 21s.

—— *Vesuvius.* 1869. Crown 8vo. 10s. 6d.

Prestwich (Joseph, M.A., F.R.S.). Geology, Chemical, Physical, and Stratigraphical. In two Volumes.

Vol. I. Chemical and Physical. Royal 8vo. 25s.
Vol. II. Stratigraphical and Physical. With a new Geographical Map of Europe. Royal 8vo. 36s.

Rolleston (George, M.D., F.R.S.). Forms of Animal Life. A Manual of Comparative Anatomy, with descriptions of selected types. Second Edition. Revised and enlarged by W. Hatchett Jackson, M.A. Medium, 8vo. cloth extra, 1l. 16s.

Smyth. A Cycle of Celestial Objects. Observed, Reduced, and Discussed by Admiral W. H. Smyth, R.N. Revised, condensed, and greatly enlarged by G. F. Chambers, F.R.A.S. 1881. 8vo. 12s.

Stewart (Balfour, LL.D., F.R.S.). An Elementary Treatise on Heat, with numerous Woodcuts and Diagrams. Fifth Edition. Extra fcap. 8vo. 7s. 6d.

Vernon-Harcourt (L. F., M.A.). A Treatise on Rivers and Canals, relating to the Control and Improvement of Rivers, and the Design, Construction, and Development of Canals. 2 vols. (Vol. I, Text. Vol. II, Plates.) 8vo. 21*s*.

—— *Harbours and Docks;* their Physical Features, History, Construction, Equipment, and Maintenance; with Statistics as to their Commercial Development. 2 vols. 8vo. 25*s*.

Walker (James, M.A.). The Theory of a Physical Balance. 8vo. stiff cover, 3*s*. 6*d*.

Watson (H. W., M.A.). A Treatise on the Kinetic Theory of Gases. 1876. 8vo. 3*s*. 6*d*.

Watson (H. W., D. Sc., F.R.S.), and S. H. Burbury, M.A.
  I. *A Treatise on the Application of Generalised Coordinates to the Kinetics of a Material System.* 1879. 8vo. 6*s*.
  II. *The Mathematical Theory of Electricity and Magnetism.* Vol. I. Electrostatics. 8vo. 10*s*. 6*d*.

Williamson (A. W., Phil. Doc., F.R.S.). Chemistry for Students. A new Edition, with Solutions. 1873. Extra fcap. 8vo. 8*s*. 6*d*.

VII. HISTORY.

Bluntschli (J. K.). The Theory of the State. By J. K. Bluntschli, late Professor of Political Sciences in the University of Heidelberg. Authorised English Translation from the Sixth German Edition. Demy 8vo. half bound, 12*s*. 6*d*.

Finlay (George, LL.D.). A History of Greece from its Conquest by the Romans to the present time, B.C. 146 to A.D. 1864. A new Edition, revised throughout, and in part re-written, with considerable additions, by the Author, and edited by H. F. Tozer, M.A. 7 vols. 8vo. 3*l*. 10*s*.

Fortescue (Sir John, Kt.). The Governance of England: otherwise called The Difference between an Absolute and a Limited Monarchy. A Revised Text. Edited, with Introduction, Notes, and Appendices, by Charles Plummer, M.A. 8vo. half bound, 12*s*. 6*d*.

Freeman (E.A., D.C.L.). A Short History of the Norman Conquest of England. Second Edition. Extra fcap. 8vo. 2*s*. 6*d*.

George (H. B., M.A.). Genealogical Tables illustrative of Modern History. Third Edition, Revised and Enlarged. Small 4to. 12*s*.

Hodgkin (T.). Italy and her Invaders. Illustrated with Plates and Maps. Vols. I—IV, A.D. 376–553. 8vo. 3*l*. 8*s*.

Hughes (Alfred). Geography for Schools. With Diagrams.
Part I. Practical Geography. Crown 8vo. 2s. 6d.
   Part II. General Geography. *In preparation.*

Kitchin (G. W., D.D.). A History of France. With numerous Maps, Plans, and Tables. In Three Volumes. *Second Edition.* Crown 8vo. each 10s. 6d.
   Vol. I. Down to the Year 1453.
Vol. II. From 1453-1624.   Vol. III. From 1624-1793.

Lucas (C. P.). Introduction to a Historical Geography of the British Colonies. With Eight Maps. Crown 8vo. 4s. 6d.

Payne (E. J., M.A.). A History of the United States of America. In the Press.

Ranke (L. von). A History of England, principally in the Seventeenth Century. Translated by Resident Members of the University of Oxford, under the superintendence of G. W. Kitchin, D.D., and C. W. Boase, M.A. 1875. 6 vols. 8vo. 3l. 3s.

Rawlinson (George, M.A.). A Manual of Ancient History. Second Edition. Demy 8vo. 14s.

Ricardo. Letters of David Ricardo to Thomas Robert Malthus (1810-1823). Edited by James Bonar, M.A. Demy 8vo. 10s. 6d.

Rogers (J. E. Thorold, M.A.). The First Nine Years of the Bank of England. 8vo. 8s. 6d.

Select Charters and other Illustrations of English Constitutional History, from the Earliest Times to the Reign of Edward I. Arranged and edited by W. Stubbs, D.D. Fifth Edition. 1883. Crown 8vo. 8s. 6d.

Stubbs (W., D.D.). The Constitutional History of England, in its Origin and Development. Library Edition. 3 vols. demy 8vo. 2l. 8s.
   Also in 3 vols. crown 8vo. price 12s. each.

—— *Seventeen Lectures on the Study of Medieval and Modern History*, &c., delivered at Oxford 1867-1884. Crown 8vo. 8s. 6d.

Wellesley. A Selection from the Despatches, Treaties, and other Papers of the Marquess Wellesley, K.G., during his Government of India. Edited by S. J. Owen, M.A. 1877. 8vo. 1l. 4s.

Wellington. A Selection from the Despatches, Treaties, and other Papers relating to India of Field-Marshal the Duke of Wellington, K.G. Edited by S. J. Owen, M.A. 1880. 8vo. 24s.

A History of British India. By S. J. Owen, M.A., Reader in Indian History in the University of Oxford. In preparation.

VIII. LAW.

Alberici Gentilis, I.C.D., I.C., De Iure Belli Libri Tres.
Edidit T. E. Holland, I.C.D. 1877. Small 4to. half morocco, 21*s.*

Anson (Sir William R., Bart., D.C.L.). Principles of the
English Law of Contract, and of Agency in its Relation to Contract. Fifth Edition. Demy 8vo. 10*s.* 6*d.*

—— *Law and Custom of the Constitution.* Part I. Parliament. Demy 8vo. 10*s.* 6*d.*

Bentham (Jeremy). An Introduction to the Principles of
Morals and Legislation. Crown 8vo. 6*s.* 6*d.*

Digby (Kenelm E., M.A.). An Introduction to the History of
the Law of Real Property. Third Edition. Demy 8vo. 10*s.* 6*d.*

Gaii Institutionum Juris Civilis Commentarii Quattuor; or,
Elements of Roman Law by Gaius. With a Translation and Commentary by Edward Poste, M.A. Second Edition. 1875. 8vo. 18*s.*

Hall (W. E., M.A.). International Law. Second Ed. 8vo. 21*s.*

Holland (T. E., D.C.L.). The Elements of Jurisprudence.
Fourth Edition. Demy 8vo. 10*s.* 6*d.*

—— *The European Concert in the Eastern Question*, a Collection of Treaties and other Public Acts. Edited, with Introductions and Notes, by Thomas Erskine Holland, D.C.L. 8vo. 12*s.* 6*d.*

Imperatoris Iustiniani Institutionum Libri Quattuor; with
Introductions, Commentary, Excursus and Translation. By J. B. Moyle, B.C.L., M.A. 2 vols. Demy 8vo. 21*s.*

Justinian, The Institutes of, edited as a recension of the
Institutes of Gaius, by Thomas Erskine Holland, D.C.L. Second Edition, 1881. Extra fcap. 8vo. 5*s.*

Justinian, Select Titles from the Digest of. By T. E. Holland,
D.C.L., and C. L. Shadwell, B.C.L. 8vo. 14*s.*

Also sold in Parts, in paper covers, as follows:—
Part I. Introductory Titles. 2*s.* 6*d.*   Part II. Family Law. 1*s.*
Part III. Property Law. 2*s.* 6*d.*   Part IV. Law of Obligations (No. 1). 3*s.* 6*d.*
Part IV. Law of Obligations (No. 2). 4*s.* 6*d.*

Lex Aquilia. The Roman Law of Damage to Property:
being a Commentary on the Title of the Digest 'Ad Legem Aquiliam' (ix. 2). With an Introduction to the Study of the Corpus Iuris Civilis. By Erwin Grueber, Dr. Jur., M.A. Demy 8vo. 10*s.* 6*d.*

Markby (W., D.C.L.). Elements of Law considered with reference to Principles of General Jurisprudence. Third Edition. Demy 8vo. 12s.6d.

Stokes (Whitley, D.C.L.). The Anglo-Indian Codes.
    Vol. I. *Substantive Law.* 8vo. 30s.
    Vol. II. *Adjective Law.* In the Press.

Twiss (Sir Travers, D.C.L.). The Law of Nations considered as Independent Political Communities.
    Part I. On the Rights and Duties of Nations in time of Peace. A new Edition, Revised and Enlarged. 1884. Demy 8vo. 15s.
    Part II. On the Rights and Duties of Nations in Time of War. Second Edition, Revised. 1875. Demy 8vo. 21s.

IX. MENTAL AND MORAL PHILOSOPHY, &c.

Bacon's Novum Organum. Edited, with English Notes, by G. W. Kitchin, D.D. 1855. 8vo. 9s. 6d.

—— Translated by G. W. Kitchin, D.D. 1855. 8vo. 9s. 6d.

Berkeley. The Works of George Berkeley, D.D., formerly Bishop of Cloyne; including many of his writings hitherto unpublished. With Prefaces, Annotations, and an Account of his Life and Philosophy, by Alexander Campbell Fraser, M.A. 4 vols. 1871. 8vo. 2l. 18s.
    The Life, Letters, &c. 1 vol. 16s.

—— *Selections from.* With an Introduction and Notes. For the use of Students in the Universities. By Alexander Campbell Fraser, LL.D. Third Edition. Crown 8vo. 7s. 6d.

Fowler (T., D.D.). The Elements of Deductive Logic, designed mainly for the use of Junior Students in the Universities. Ninth Edition, with a Collection of Examples. Extra fcap. 8vo. 3s. 6d.

—— *The Elements of Inductive Logic*, designed mainly for the use of Students in the Universities. Fourth Edition. Extra fcap. 8vo. 6s.

—— *and Wilson (J. M., B.D.). The Principles of Morals* (Introductory Chapters). 8vo. *boards*, 3s. 6d.

—— *The Principles of Morals.* Part II. (Being the Body of the Work.) 8vo. 10s. 6d.

Edited by T. FOWLER, D.D.

Bacon. Novum Organum. With Introduction, Notes, &c. 1878. 8vo. 14s.

Locke's Conduct of the Understanding. Second Edition. Extra fcap. 8vo. 2s.

Danson (J. T.). The Wealth of Households. Crown 8vo. 5*s*.

Green (T. H., M.A.). Prolegomena to Ethics. Edited by A. C. Bradley, M.A. Demy 8vo. 12*s*. 6*d*.

Hegel. The Logic of Hegel; translated from the Encyclopaedia of the Philosophical Sciences. With Prolegomena by William Wallace, M.A. 1874. 8vo. 14*s*.

Hume's Treatise of Human Nature; reprinted from the Original Edition, and edited by L. A. Selby-Bigge, M.A., Fellow and Lecturer of University College. Crown 8vo. 9*s*.

Lotze's Logic, in Three Books; of Thought, of Investigation, and of Knowledge. English Translation: Edited by B. Bosanquet, M.A., Fellow of University College, Oxford. Second Edition. 2 vols. Crown 8vo. 12*s*.

—— *Metaphysic,* in Three Books; Ontology, Cosmology, and Psychology. English Translation: Edited by B. Bosanquet, M.A. Second Edition. 2 vols. Crown 8vo. 12*s*.

Martineau (James, D.D.). Types of Ethical Theory. Second Edition. 2 vols. Crown 8vo. 15*s*.

—— *A Study of Religion: its Sources and Contents.* 2 vols. 8vo. 24*s*.

Rogers (J. E. Thorold, M.A.). A Manual of Political Economy, for the use of Schools. Third Edition. Extra fcap. 8vo. 4*s*. 6*d*.

Smith's Wealth of Nations. A new Edition, with Notes, by J. E. Thorold Rogers, M.A. 2 vols. 8vo. 1880. 21*s*.

X. FINE ART.

Butler (A. J., M.A., F.S.A.) The Ancient Coptic Churches of Egypt. 2 vols. 8vo. 30*s*.

Head (Barclay V.). Historia Numorum. A Manual of Greek Numismatics. Royal 8vo. *half morocco,* 42*s*.

Hullah (John). The Cultivation of the Speaking Voice. Second Edition. Extra fcap. 8vo. 2*s*. 6*d*.

Jackson (T. G., M.A.). Dalmatia, the Quarnero and Istria; with Cettigne in Montenegro and the Island of Grado. By T. G. Jackson, M.A., Author of 'Modern Gothic Architecture.' In 3 vols. 8vo. With many Plates and Illustrations. *Half bound,* 42*s*.

Ouseley (Sir F. A. Gore, Bart.). A Treatise on Harmony.
Third Edition. 4to. 10*s*.

—— *A Treatise on Counterpoint, Canon, and Fugue,* based upon that of Cherubini. Second Edition. 4to. 16*s*.

—— *A Treatise on Musical Form and General Composition.* Second Edition. 4to. 10*s*.

Robinson (J. C., F.S.A.). A Critical Account of the Drawings by Michel Angelo and Raffaello in the University Galleries, Oxford. 1870. Crown 8vo. 4*s*.

Troutbeck (J., M.A.) and R. F. Dale, M.A. A Music Primer (for Schools). Second Edition. Crown 8vo. 1*s*. 6*d*.

Tyrwhitt (R. St. J., M.A.). A Handbook of Pictorial Art. With coloured Illustrations, Photographs, and a chapter on Perspective by A. Macdonald. Second Edition. 1875. 8vo. half morocco, 18*s*.

Upcott (L. E., M.A.). An Introduction to Greek Sculpture. Crown 8vo. 4*s*. 6*d*.

Vaux (W. S. W., M.A.). Catalogue of the Castellani Collection of Antiquities in the University Galleries, Oxford. Crown 8vo. 1*s*.

The Oxford Bible for Teachers, containing Supplementary HELPS TO THE STUDY OF THE BIBLE, including Summaries of the several Books, with copious Explanatory Notes and Tables illustrative of Scripture History and the characteristics of Bible Lands; with a complete Index of Subjects, a Concordance, a Dictionary of Proper Names, and a series of Maps. Prices in various sizes and bindings from 3*s*. to 2*l*. 5*s*.

Helps to the Study of the Bible, taken from the OXFORD BIBLE FOR TEACHERS, comprising Summaries of the several Books, with copious Explanatory Notes and Tables illustrative of Scripture History and the Characteristics of Bible Lands; with a complete Index of Subjects, a Concordance, a Dictionary of Proper Names, and a series of Maps. Crown 8vo. *cloth,* 3*s*. 6*d*.; 16mo. *cloth,* 1*s*.

LONDON: HENRY FROWDE,
OXFORD UNIVERSITY PRESS WAREHOUSE, AMEN CORNER,

OXFORD: CLARENDON PRESS DEPOSITORY,
116 HIGH STREET.

☞ *The* DELEGATES OF THE PRESS *invite suggestions and advice from all persons interested in education; and will be thankful for hints, &c. addressed to the* SECRETARY TO THE DELEGATES, *Clarendon Press, Oxford.*

www.ingramcontent.com/pod-product-compliance
Lightning Source LLC
Chambersburg PA
CBHW030011240426
43672CB00007B/903